A
Roadside Guide
to the
Geology
of the
Great Smoky Mountains
National Park

A
Roadside Guide
to the
Geology
of the
Great Smoky Mountains
National Park

Harry L. Moore

The University of Tennessee Press / Knoxville

Cloth: 1st printing, 1988.
Paper: 1st printing, 1988; 2nd printing, 1992; 3rd printing, 1995;
           4th printing, 2004.

The paper in this book meets the minimum requirements of the
American National Standard for Permanence of Paper for Printed
Library Materials. ∞   The binding materials have been chosen for
strength and durability.

Frontispiece: Looking north toward Chilhowee Mountain from the
Foothills Parkway (beginning of Road Log 5). All photographs by
the author.

Library of Congress Catologing-in-Publication Data

Moore, Harry L., 1949–

    A roadside guide to the geology of the Great Smoky Mountains
National Park / Harry L. Moore.—1st ed.
    p.   cm.
    Bibiography: p.
    Includes Index.
    ISBN 0-87049-558-5 (pbk. alk. paper)
    1. Geology—Great Smoky Mountians National Park (N.C. and
Tenn.)—Guide-books. 2. Great Smoky Mountains National Park
(N.C. and Tenn.)—Description and travel—Guide-books. I. Title
QE78.5.M66   1988
917.688'90453—dc19   87-18796   CIP

For Alice Ann

Acknowledgments

During the course of my research and preparation of the manuscript, a number of people have offered me encouragement and assistance.

Don Byerly, Professor of Geology at the University of Tennessee, Knoxville, and Stuart Maher, retired geologist with the Tennessee Division of Geology, kindly read the manuscript and offered many suggestions which have improved both its accuracy and its style.

Michael Clark and Paul Delcourt, both Professors of Geology at the University of Tennessee, gave me particular help with the text on Cenozoic deposits.

The Great Smoky Mountains Natural History Association provided the necessary map information for road log details, and the National Park Service staff at the Great Smoky Mountains National Park offered assistance in the complicated realm of Park information.

Thanks also go to Judy Sayne, who typed the many preliminary drafts of the manuscript, and Jeff Snyder, who drafted the series of schematic diagrams on wedge failure.

I am deeply indebted to my patient editor, Bettie McDavid Mason, who spent untold hours struggling with an unruly manuscript. I would also like to express my thanks to Carol Orr and the staff of the University of Tennessee Press for their advice and assistance in the production of this book.

As always, I am grateful to my wife, Alice Ann Moore. Without her patience and understanding, I could never have undertaken this project.

Contents

Illustrations

Maps

Figures

Tables

Photographs

PART 1

**Of Time,
Rocks,
and Geologic
Architecture**

Alum Cave Bluffs on Mt. Le Conte, outcroppings of the Anakeesta Formation, get their name from mineral residues found in the bedrock.

Introduction

Wherever you look, the Great Smoky Mountains National Park offers you hints of its underlying geology. Behind and below the beauty and splendor of the mountains is a history of geologic upheaval. Tremendous pressures have formed rocks and minerals, oceans have deposited sediments, and wind and water have eroded the surface to form the fabric of the mountains.

Even if you are only a casual observer, you can hardly miss the effects of the Park's geology. Beside the road, mountain streams like the Little Pigeon rush around massive boulders and play across smaller stones and gravel. High above, striking prominences like the Chimney Tops and Charlies Bunion punctuate the skyline. And amid some of the highest mountains in the Eastern United States, Cades Cove and others like it open out as wide, almost level valleys. With this book in hand, you can have a guided tour of some of the most interesting features of the Park—and gain an understanding of how they came to be.

What you see in the Park today is, of course, a result of its geologic history. For a student of geology, the history of the Park is not just the little more than half a century since these mountains were placed under Federal protection. In-

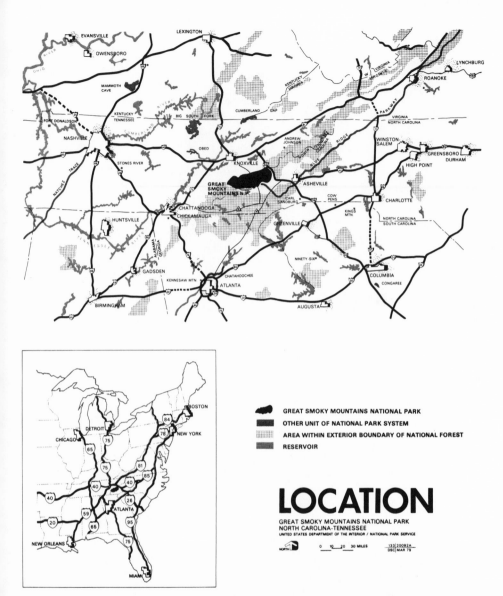

Map 1. *General location map of the Great Smoky Mountains National Park (from National Park Service, Draft Environmental Impact Statement for the General Management Plan, 1979).*

stead, it is ten million times that long, for the oldest rocks in these mountains were formed more than half a billion years ago!

You do not need to be a professional geologist to understand the long history of these mountains and how they came to appear as you see them today. But it does help to understand a few basic concepts. In diagrams and words, this book will introduce you to the formation of the rocks in the sediments below ancient oceans and show you how these rocks were pushed up, folded, and fractured to build the structures of the mountains. You will also learn how, after this long history, geologic forces are still reshaping the Smokies.

To appreciate the geology of the Park to the fullest, you need to go out into the mountains themselves. To lead you to some of the most interesting features around you, this book provides annotated road logs for five tours through the Park and along its boundaries. The routes lead to some of the most popular—and geologically interesting—sites in the Park, including Clingmans Dome and Cades Cove. (For an even closer look at the geology of the Park, you may wish to walk the optional hiking trail provided with each road route.) You will find many photographs to help you identify geologic phenomena along the tours, particularly where the lush vegetation of the mountains hides the underlying features.

Along with this book you will need a good map of the Park. Maps are available at the Oconaluftee and Sugarlands Visitor Centers in the Park. For identifying mountains and other geologic structures, one of the best maps is that prepared by the U.S. Geologic Survey in 1949 and revised periodically since then (N3515–W8300).

You can enjoy the geology of the Smokies throughout the year. Unlike seasonal plants and migratory birds, the rocks and minerals are always there. In fact, the changing of the seasons serves to show off one feature and then another.

The unique geology of the Great Smoky Mountains has produced a unique ecosystem, with more than 2000 species of plants and over 400 species of animals. This system is fragile, however, and you should not collect either rocks or living organisms in the Park. Study and enjoy what you find where you find it, but leave it in place so that this jewel of nature that is the Great Smoky Mountains National Park can last for many years to come.

Fluffy cumulus clouds frame Rich Mountain along the northwest boundary of the Great Smoky Mountains National Park. Tuckaleechee Cove (Townsend, TN) is located in the valley below. A similar cove is Cades Cove, located on the other side of Rich Mountain.

Mt. Le Conte (seen from the Gatlinburg Bypass) wears a blanket of late winter snow.

Vegetation, masking much of the bedrock in the Smokies, makes exposures of rock strata somewhat less visible. In the higher elevations of the Park, evergreens such as red spruce and Fraser fir flourish much like their counterparts in southern Canada.

Massive exposures of Thunderhead Sandstone are found along U.S. 441 near the Chimney Tops area. The Anakeesta Formation, consisting of metamorphic rocks such as slate, phyllite, and metasilt-stone, forms the crest of the Chimney Tops (upper right-hand part of photograph).

Rhododendron, hemlock, and hardwoods compose much of the vegetative growth in the lower portions of the Smokies.

A popular destination for hikers is Laurel Falls, formed where a small tributary stream of the Little River flows over massive exposures of Thunderhead Sandstone.

The foothills of the Smokies are underlain by somewhat less resistant rock strata which have weathered to produce the rolling topography shown here. The view is from Bull Head on Mt. Le Conte toward Sevierville, TN, with the less mountainous Ridge and Valley province in the distance.

Rich Mountain, framed by early morning cumulus clouds and rising fog.

Many varieties of ferns thrive in the Smokies because of the humid climate and the rich soils derived from weathering of the underlying bedrock.

Charlies Bunion, a barren rock prominence located along the crest of the Smokies about 4 miles east of Newfound Gap, is composed of strata of the Anakeesta Formation. The Appalachian Trail passes over the rock pinnacle.

The center of the Park is composed of high topographic features such as Mt. Le Conte, Newfound Gap, and Clingmans Dome, seen in this view from Cove Mountain at the northern Park boundary, just outside Gatlinburg, TN.

Table 1. *Geologic Time Scale*

Duration in Millions of Years	Millions of Years Before the Present	Time Units			Rock Units	Geologic Events	Notable Topographic Features in Park
		Era	Periods	Epochs			
1.8	0.01	Cenozoic	Quaternary	Holocene	Alluvium	Stream erosion, mass wasting	Stream beds, flood plains
				– – – –			
				Pleistocene	Talus slopes, block fields	Increased frost action, weathering, stream erosion during climatic fluctuations	Valley floors, mountain slopes
	1.8				Alluvium		Sinkholes, caves
63.2	65		Tertiary (or Paleogene/Neogene)			Long erosion, establishment of modern drainage patterns Sporadic uplift	
165		Meso-zoic	Cretaceous			Prolonged erosion	
			Jurassic			Uplift due to isostatic adjustments	
	230		Triassic				
		Paleozoic	Permian			Gatlinburg faults (?)	
			Pennsylvanian			Great Smoky thrust faults	
						Allegheny orogeny, folding, regional metamorphism	
			Mississippian		(outside Park)	Deposition in vicinity of Great Smoky Mountains	
			Devonian				
340			Silurian			Erosional interval	
			Ordovician	(middle)	(outside Park)	Deposition in vicinity of Great Smoky Mountain Park	
						Taconic orogeny, folding Greenbrier thrust faults	
						Unconformity	
				(lower)	Knox Group	Carbonates (exposed in coves)	Cades Cove, Whiteoak Sink
			Cambrian	(lower)	Chilhowee Group	Thick deposits, mostly sands, on continental shelf	Chilhowee Mountain
						Exposed in valleys and ridges north and west of Park	Green Mountain
	570		?			Disconformity (?)	
4000		Precambrian	Proterozoic Z (late Precambrian)		Ocoee Supergroup — Walden Creek Group	Slates, shales, siltstones	Foothills of Smokies
					Unclassified formations	Sandstones	
					Great Smoky Group	Mainly thick-bedded sandstones	Crest of Smokies (Chimneys, Mt. Le Conte, Clingmans Dome)
					Snowbird Group	Sandstones, siltstones	Foothills of Smokies
						Erosional interval	
	4600		Proterozoic Y		Basement complex	Accumulated sediments and intrusives now highly metamorphosed and deformed	Small ridge in Southeast part of Smokies; Hyatt Ridge

Note: Only Precambrian and Paleozoic time are represented in the rock strata of the Great Smoky Mountains National Park; however, some stream sediments and colluvium of recent epochs can be found throughout the Park.

Geologic Terms

Discussing the history of the Great Smoky Mountains will be made simpler by introducing a few key geological terms.

Time

Geologists divide time into long periods called *eras*. The four main eras, listed from youngest to oldest, are *Cenozoic, Mesozoic, Paleozoic,* and *Precambrian*.

Geologists know that early in our earth's history life was not present. But about one billion years ago life forms began to develop. This earliest time span—from the earth's beginning, about 4.5 billion years ago, to about 500 million years ago—is known as the Precambrian Era. There is fossil evidence that late in Precambrian time soft-shelled primitive life forms existed. A majority of the rocks in the Great Smoky Mountains are believed to be of late Precambrian age.

The next oldest era is the Paleozoic (meaning "ancient life"), which is subdivided into seven separate systems. Of the rocks found in the Park, however, the Paleozoic ones comprise only about 10 percent and are limited to small areas, e.g. Cades Cove, Chilhowee Mountain, and Green Mountain.

In this geologic time sequence, the next-to-youngest era is the Mesozoic ("middle life"), often called the Age of Reptiles because dinosaurs ruled the earth then. Mesozoic age rocks are not found in the Great Smoky Mountains National Park, however.

The most recent era—the one in which we live—is the Cenozoic ("recent life"), also known as the Age of Mammals. In the Smokies, Cenozoic age rocks are generally confined to colluvium, talus, and other recent unconsolidated surface deposits.

Structures

In discussing the geology of the Smokies, some technical terms must be used to express the character of the *bedrock*, the solid rock exposed at the surface of the earth or overlain by unconsolidated material such as soil or colluvium.

Rock units of sufficient size to be mapped on a scale such as the 1″ = 2,000′ standard employed in the United States Geologic Survey 7½ minute quadrangle maps are generally divided into *groups* or *formations* (for example, the Great Smoky Group or the Thunderhead Formation). A formation name is simply a convenient means by which geologists can refer to a group of rocks which have like characteristics and can be mapped at the earth's surface. Usually a formation is named for the area (or geographic feature, such as a river) in or near which it characteristically appears; thus Cades Sandstone for Cades Cove, Metcalf Phyllite for Metcalf Bottoms, Pigeon Siltstone for the Little Pigeon River. This is not to say that these formations are not found in other areas as well.

The kind of rocks in a given formation are usually the same, though these rock units can range in type from one place to another. For example, the Thunderhead Formation is predominantly a sandstone but ranges from a sandstone to a conglomerate (containing pebbles), a siltstone, or a shale—all within a half mile or less! The color of a rock may range as well; indeed, color is only site specific for rocks and should not be used as a general guide for identifying formations.

A *fault* is a break in a rock mass along which movement has taken place. A *thrust fault* is formed by one rock mass

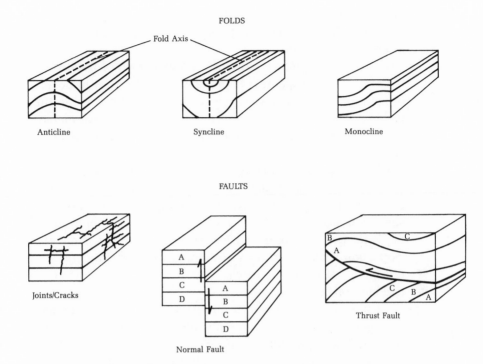

FOLDS

Fold Axis

Anticline   Syncline   Monocline

FAULTS

Joints/Cracks

Normal Fault

Thrust Fault

Figure 1. *Geologic rock structures.*

being pushed over another body of rocks at a low angle; generally the overriding mass is older than the overridden one. The majority of faults in the Park are thrust faults.

While observing the rocks in the Park, you can easily see that the rock strata fracture and break along definite planes. In fact, these planes of broken rock dominate most rock exposures in the Park. The fractures also provide key avenues for breaking down and eroding the rock. Geologists have recognized two major kinds of fracture in the rocks of the Smokies: jointing and cleavage.

A simple breaking pattern in which fractures are more or less vertical to bedding and have shown no appreciable movement is known as a *joint*. Although most of the rocks in the Park exhibit some degree of jointing, the sandstone beds of the Hesse Formation along the Foothills Parkway on Chil-

howee Mountain (Road Log 5) and the Cades Sandstone at
Abrams Falls (Road Log 4, Hiking Trail) are particularly good
examples of the pattern.

A second and somewhat more complex form of fracturing
is known as *cleavage:* this is the tendency of a rock to break
or split along closely spaced planes which are inclined to the
bedding planes of strata. In an outcrop these cleavage planes
may appear as cracks that tend to develop a diagonal pattern
between two bedding planes.

Cleavage generally develops in response to pressure and is
often associated with metamorphism, faulting, and folding of
rocks. Several types of cleavage—including fracture cleavage,
shear cleavage, and slaty cleavage—occur in rocks found in
the Park. Determination of cleavage types, however, is depen-
dent upon such factors as surrounding rock structure, arrangement of mineral particles within the rock, and the actual mineral content of the rock.

Of the rock types found in the Park, those exhibiting good
cleavage development include phyllite, slate, and slaty metasiltstone units. Metcalf Phyllite (Road Log 4), Anakeesta Formation (Road Logs 2 and 3), and Pigeon Siltstone (Road Log
1) all characteristically develop cleavage. Observed in an outcrop or roadcut, the broken pieces of rock exhibiting cleavage usually appear as polygonal blocks whose sides are not
at right angles to each other.

During the complex geologic history of the Park, numerous layers of rock have undergone multiple episodes of deformation. This process often results in the development of several sets of cleavage planes within a given rock mass. The
Metcalf Phyllite along Laurel Creek Road (Road Log 4) provides particularly good examples of multiple cleavage.

Rock strata in the Smokies are commonly folded into several structural configurations. An *anticline* is an upward bend
or arch in rocks, easily observed in layered rock strata. A
syncline is the opposite of an anticline, that is, a downward
fold or bend in the rock strata. Variations that occur in these
two major types of folds include *asymmetrical* folds, *mono-
clines,* or *chevron* folds.

Kinds of Rocks

The three basic kinds of rocks on earth—igneous, sedimentary, and metamorphic—all occur in the Great Smoky Mountains.

Igneous rocks originate from molten rock (magma) in the earth's interior. Igneous rocks may be extrusive (coming to the surface) or intrusive (that is, the molten material rises but cools slowly below the surface). Intrusive rocks are generally coarse-grained and extrusive igneous rocks fine-grained, as the rate of cooling governs crystal size.

Sedimentary rocks are formed from sediments deposited on land, in seas, or along and in streams and lakes. Although the majority of rocks in the Smokies are metamorphosed sedimentary rocks, some unaltered sedimentary rocks also occur: these are limestones and shales found primarily in Cades Cove. As can be logically deduced, the older sedimentary rocks are on the bottom of a section and the younger ones are at the top in undisturbed areas. Sedimentary rocks exhibit bedding: that is, they are layered or stratified. Most fossils are found in sedimentary rocks.

Metamorphic rocks are preexisting rock types (originally either igneous or sedimentary) that have been altered by heat and/or pressure and may be chemically altered as well. The alteration may range from only slightly metamorphosed to highly metamorphosed. Most of the rocks in the Great Smoky Mountains are metamorphosed sedimentary rocks. Again the metamorphism ranges in degree, but most rocks in the Park are lightly to moderately metamorphosed, and, accordingly, sedimentary rock features are still identifiable (bedding or layers of strata are distinguishable).

Minerals

Rocks are composed of minerals, generally in mixtures. *Minerals* are naturally occurring substances that have definite chemical compositions and atomic arrangement and usually specific crystal forms. As rocks are mixtures of minerals, their compositions are not fixed.

Mineral composition and grain size determine rock types. Thus, an igneous rock with readily visible mineral grains consisting chiefly of quartz and feldspar (and other minerals in lesser quantities) is a *granite*. A sedimentary rock chiefly composed of calcite is *limestone*. If clay predominates, the rock is a *shale*. Quartz grains predominate in *quartz sandstones*. If metamorphosed, granites form *gneisses* or *schists*, limestones become *marbles*, and shales become *slates*.

Following are some minerals commonly occurring in the Smokies:

Quartz is a clear, hard, shiny to glassy mineral that cannot be scratched with a knife. It is the primary mineral in most sandstones and granites. Several varieties occur in the Park, including milky quartz (milky-white), rose quartz (pink), and smoky quartz (gray to clear).

Calcite is a white to light-gray mineral composing most limestones and marbles. It can be scratched with a knife but not with a fingernail.

Mica is a shiny, soft, black or silvery-white mineral that occurs as plates or layers of material. Mica is very common in granite and schist. Black mica is called biotite; white mica is muscovite.

Pyrite, commonly called fool's gold, is a brittle, metallic, brass-colored compound of iron and sulfur. It may occur as large crystal cubes one-half inch in width or very fine grains barely visible to the naked eye. This mineral is very common in the Anakeesta Formation in the Park.

Limonite is a rusty, yellowish-brown, or buff-colored compound occurring as tiny grains that give an overall yellowish appearance. In the Smokies limonite is often found as a result of the weathering of pyrite. It was formerly used in iron-making.

Feldspar is a white to pale-yellow to pinkish-gray mineral composed of alumino-silicates and soda-lime silicates. Feldspar is commonly found in Park rocks such as sandstones, conglomeratic sandstones, and graywacke (a sedimentary rock composed chiefly of quartz, feldspar, and other rock fragments).

Kaolinite is a white to tan silicate clay mineral usually occurring in colloidal masses. Kaolinite, a secondary mineral resulting from the weathering of feldspar, can be found in the Park as clay deposits within the soil.

Numerous other minerals may be found in the Park and easily identified with the use of proper manuals and techniques. (For a list of the more common rock units and types, see Table 2.)

This horizontal layering is the original stratification of the bedrock composing the Anakeesta Formation.

A View of Geologic History

Geologists recognize that the earth has a very long history and during its existence has undergone great changes. Such changes continue. Lands and seas change their relative locations by plate movement. Mountains are raised, only to be worn away by erosion. Streams cut deep canyons and build large deltas at their mouths. Erupting volcanoes flood areas with lava flows and ash. Earthquakes shake and vibrate land surfaces. The earth is dynamic and everchanging: today's scenery is not the same as yesterday's, and tomorrow's will be different still.

You can think of scenery as the result of processes such as erosion, mountain-building, and vulcanism acting on the materials composing the earth. As the process and/or materials are changed, so are the resulting landscapes. Generally these changes are gradual enough to be almost imperceptible in a man's life span, but over long periods of time the effects are enormous.

Research over the past half-century has demonstrated that the earth consists of concentric layers. The lightest outer layer, the crust, consists of continental rocks (least dense) and ocean basin rocks. Both of these rock types float on a denser, probably plastic layer called the asthenosphere (the

The crest of the Smokies is composed of very resistant metamor-phosed sedimentary rocks which have been folded, faulted, and subjected to many thousands of years of weathering.

"without strength" zone). Heat, generated by enormous pressures and the decay of radioactive minerals, migrates through the asthenosphere in slow-moving giant convection currents toward the surface. As the currents are cooled by heat loss, they descend once again to be heated and then migrate to other locations. By such forces the continents and ocean basins are stressed and broken into plates that are propelled across the asthenosphere like blocks of wood on hot water.

These drifting plates collide, sideswipe one another, but continue in constant motion, albeit very, very slowly. Thus ocean basins open and close, and continents collide and buckle along their boundaries or are dragged down along descending convection currents (trenches). Ascending hot currents cause spreading of the rocks (rifts) along which volcanoes are formed and earthquakes occur (earthquakes also frequent descending zones).

If you could view the earth as it was some 500 to 600 million years ago, you would see the Western Hemisphere in collision with the Eastern Hemisphere. Initially the hemispheres were separated by an ancestral Atlantic Ocean, which was then a long, narrow seaway, not a broad expanse of water.

The collision of the continental masses now composing the Eastern and Western Hemispheres compressed the rocks, cast them into a series of folds extending northeasterly to southwesterly, and slid rocks across other rocks in great flat thrust plates. Some geologists think large fragments of the Eastern Hemisphere were left behind in North America when the continents pulled apart after their collision. Such "exotic terrains" (detached remnants of continents) explain similarities between rock masses now separated by the Atlantic Ocean.

The geologic history of the Smokies began hundreds of millions of years ago as the earth was forming and reshaping its surfaces. While the landmass continents were moving about the surface of the earth, ocean sediments being deposited along the margins of the continents (and the oceans in between) were to become the fabric of the bedrock that composes such peaks as Mt. Le Conte, Clingmans Dome, Mt. Cammerer, and Mt. Guyot.

This geologic history also involves millions of years of erosional forces carving and eating away the bedrock. Snow, rain, ice, and wind have all acted upon the earth's surface in shaping and forming the topography of the Great Smoky Mountains. In more recent times flash floods, causing debris

The rock strata which have been folded and faulted are also extremely fractured, as in this outcrop of jointed Thunderhead Sandstone in the Elkmont section of the Park.

flows and massive landslides, have made their mark on developing what we see today. In addition, the day-by-day, ongoing processes of erosion continue to reshape and redefine the mountain peaks, ravines, ridges, and valleys.

All of these geologic processes began many millions of years ago. Millions of years!! Just what is a million? One million seconds are equal to 16,666.6 minutes, which are equal to 277.7 hours, which are equal to 11.6 days. One million days are equal to 2,739.7 years, a span greater than that between the present time and the Golden Age of ancient Greece.

When we speak in terms of geologic time, the word million is always there! One million years, however, is just a blip of time compared to the age of the earth, estimated to be near 4.5 *billion* years.

The rocks that compose the Greak Smoky Mountains range in age from over 1 billion years to near 350 million. Such time spans are almost incomprehensible to us, with our mere 75 years or so of life expectancy.

Just how did these rocks get here, and how did these mountains form?

The majority of the rock strata found in the Park are metamorphosed sedimentary rocks, such as this metasiltstone.

*Cleavage fracture character-
izes the thin-bedded and
more shaly or slaty rock
units in the Smokies. Cleavage fracture results from
pressure and/or stress induced on the bedrock and
develops in planes which are
not perpendicular to the bedding of the strata.*

The Great Smoky Mountains are in an area designated by
geomorphologists (scientists who study the earth's land forms)
as the Blue Ridge province. So called because of their characteristic bluish haze, this range of mountains stretches from
northern Georgia to Maine. The Great Smoky Mountains are
one of the large mountain groups which form the Blue Ridge
province. To the west of the Smokies is the Ridge and Valley
province, with its characteristic landscape of parallel ridges
and valleys. To the east of the Blue Ridge is the Piedmont
province, a low area with moderately hilly to rolling land.

The rocks that compose the Smokies originated in a sea or
ocean which was located between several large drifting continents, of which North America and Africa are but two.
That the ocean was for the most part a moderately shallow
sea platform is reflected by the rock types found in the Smokies
(limestones, sandstones, shales).

As the sand and clay particles were accumulating on the
sea floor and the calcium carbonate muds (later to be formed
into limestones) were developing along the ocean platforms

The rock strata found in the Smokies have been folded into many positions. These are vertical joints in tilted sandstone which was originally deposited in a horizontal position (Cades Sandstone near Abrams Falls).

of the drifting continents, the sediments were being readied for the squeeze. The moving continents, drawn by thermal convection currents deep within the earth's mantle, continued to move closer and closer. Deep ocean trenches were sources of subducting zones, where the earth's crust was drawn back down into the mantle. As the crust moved down, some continents were dragged along for the ride. The movement of the continents, occurring over millions of years, eventually resulted in a catastrophic collision. Where the continents collided, huge mountain ranges formed: for example, the Himalayas developed where the Indian continent collided with the Siberian plate of the Euroasian continent.

The collision of continents which formed the Smokies is known among geological scientists as the Appalachian orogeny (mountain-building episode). This event took place approximately 200 million years ago at the end of the Paleozoic time.

As the continents collided, all the rock strata and ocean sediments that were located between the moving continents were crushed, broken, folded, and eventually faulted (broken

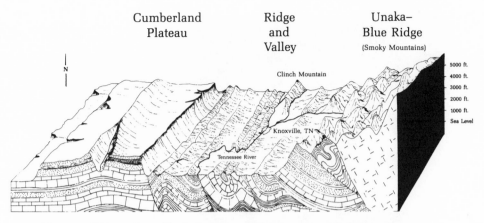

Cumberland
Plateau

Ridge
and
Valley

Unaka–
Blue Ridge
(Smoky Mountains)

Clinch Mountain

Knoxville, TN

Tennessee River

5000 ft.
4000 ft.
3000 ft.
2000 ft.
1000 ft.
Sea Level

Figure 2. *This geologic block diagram of East Tennessee illustrates the complex structural arrangement of the bedrock found in the Great Smoky Mountains (Blue Ridge province) and the adjacent Ridge and Valley province in contrast to the relatively simple, flat-lying character of the bedrock along the Cumberland Plateau (H. Moore, 1978).*

and shoved over each other). Rock strata which had once been horizontal were tilted at steep angles, folded, fractured, and exposed to extreme pressures and heat generated from the colliding continents. As a result, the rock strata have been metamorphosed (changed to varying degrees from the original material). These metamorphic rocks are classified as metasandstones, metasiltstones, slates, phyllites, and quartzites; all of these are found in the Park. The rock strata that were folded and faulted were uplifted, forming a new landmass known as the Appalachian Mountains, the Blue Ridge Mountains, and the Great Smoky Mountains.

Through the 200 million years since the building of the Appalachian Mountains, the rock strata have been exposed to weathering processes such as rain, sleet, snow, ice, heat, and wind. The rock strata least resistant to weathering—such as the shales, slates, phyllites, and limestones—have been eroded away at a faster rate than the more durable rock strata like the sandstones, siltstones, quartzites, and their metamorphic equivalents.

But to return to the Appalachian orogeny. When the rock strata were being crushed by the colliding continents, large,

These snowcapped meta-sandstone boulders in the West Prong of the Little Pigeon River attest to the power of water to move mountains oceanward by way of stream sediment.

In addition to the rain and wind, ice contributes to the weathering process by prying and breaking loose pieces of bedrock during the winter months.

wide-spread thrust faults developed. One of these, known as the Great Smoky fault, moved thousands of feet of later Pre-cambrian bedrock up and over the younger Paleozoic rocks characteristic of the Ridge and Valley in East Tennessee. The older, more resistant metamorphosed rock strata formed the bulk of the Smokies we see today.

Coves and Caves

In several areas erosion has weathered through the older Pre-cambrian rocks, exposing the younger Paleozoic rocks beneath the thrust fault. Areas where the Paleozoic limestones and shales have been exposed have formed mountain coves such as Cades Cove and Wears Cove.

These coves are surrounded by older and topographically higher rock strata to form "windows" down into the younger (limestone) strata of the cove. The fertile, deep soils produced by the weathering of the limestone made the coves attractive places for pioneers to settle and farm.

In these limestone cove areas, erosion of the bedrock has produced karst features such as sinkholes and caves. Chemical weathering of the limestone bedrock—groundwater percolating down through the rock—has enlarged fractures in the rock mass to produce a number of interconnecting solution channels, forming caves. (At the surface the cavernous bedrock is reflected as sinkholes and depressions.)

At least four limestone caves are found in the Park: Gregorys Cave (in Cades Cove), Bull Cave (near Cades Cove), and Blowing Cave and Rainbow Cave (both in Whiteoak Sink). (Note: You must obtain a permit from the Superintendent's Office of the Great Smoky Mountains National Park before entering any cave in the Park.)

Bedrocks

The geology of the Great Smoky Mountains National Park is complex, due to age (most of the rocks are over a billion years old), folding, fracturing, faulting, metamorphism, and weathering. Understanding what you see at the surface is

Karst features such as caves and sinkholes are common in the lime-stone coves of the Smokies. The two people in this picture are dwarfed by the limestone bluff (60 to 80 feet high) that rims a portion of Whiteoak Sink in the northwestern part of the Park (Road Log 4). The flat land, with the highly fertile soil of the sinkhole floor, supported a small pioneer farm before the founding of the Great Smoky Mountains National Park.

A small stream flows into Whiteoak Sink over a ledge of rock to form a 30-foot-high waterfall which empties into a limestone cave. The Great Smoky thrust fault is exposed behind this water-fall: the Cades Sandstone (dark upper rock) has been thrust over the younger lime-stone (light-colored lower rock); the fault trace is al-most horizontal.

made even more difficult by the very dense vegetative cover which obscures much of the bedrock in the Park.

Most bedrocks of the Park can be divided into three basic groups: metamorphic Precambrian basement complex, slightly metamorphosed sedimentary late Precambrian rocks of the Ocoee Supergroup, and sedimentary rocks of the Appalachian Ridge and Valley.

Class I. The metamorphic rocks of the Precambrian basement complex (more than one billion years old) consist chiefly of gneisses, schists, and some granitic rocks. The basement complex forms the ancient crystalline foundation on which all the other strata of the area have been deposited. Rocks of the Precambrian basement complex are found in the southern and southeastern portions of the Park (North Carolina side).

Class II. Predominant in the Park are the metamorphosed, sedimentary late Precambrian rocks of the Ocoee Supergroup, from about 500 million to 1 billion years old. These rocks range in their degree of metamorphism from intensely metamorphosed phyllites, schists, and quartzites in the southeastern part of the Park to less metamorphosed rocks such as slates, shales, sandstones, and metasiltstones found in the northwestern part. The Thunderhead and Anakeesta formations, the two most abundant formations of the Ocoee Supergroup found in the Park, make up the bulk of the crest of the Smokies. The Ocoee Supergroup is subdivided into three main groups: the Snowbird, Great Smoky, and Walden Creek Groups. There are a total of thirteen formations in these three groups and an additional four formations not assigned to any of the groups (Table 2).

Class III. The sedimentary rocks of the Appalachian Ridge and Valley are the youngest of the rocks in the Park. Ranging in age from 300 to 500 million years, they were deposited during the Paleozoic Era. Rock types from this group include limestones, dolostones, shales, and sandstones. Surface exposures are of very limited extent in the Park and can be found only in Cades Cove, Crib Gap, Whiteoak Sink, and portions of the Foothills Parkway on Chilhowee Mountain at

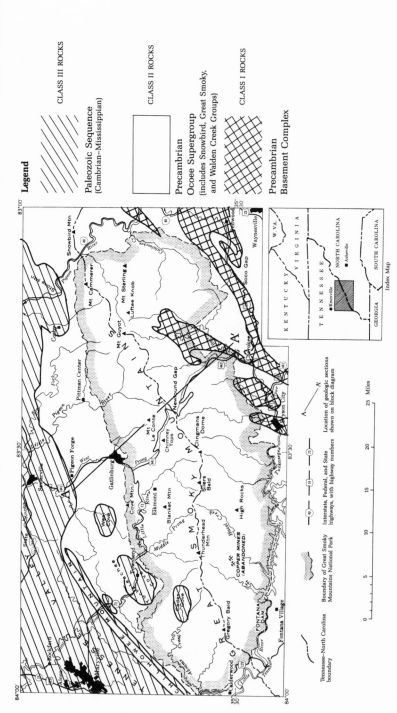

Legend

CLASS III ROCKS

Paleozoic Sequence
(Cambrian–Mississippian)

CLASS II ROCKS

Precambrian
Ocoee Supergroup
(includes Snowbird, Great Smoky,
and Walden Creek Groups)

CLASS I ROCKS

Precambrian
Basement Complex

Map 2. *A generalized geologic map of the Great Smoky Mountains National Park and vicinity shows the distribution of the three main groups of rocks found in the Park: Class I, Precambrian Basement Complex (granites and gneiss); Class II, Precambrian Ocoee Supergroup (metamorphosed sedimentary rocks, slates, metasandstones, etc.); Class III, Paleozoic Sqeuence (sedimentary strata including limestone and shale). Base map after P. B. King, 1949—on back of U.S.G.S. map, "The Great Smoky Mountains National Park and Vicinity" (revised 1972).*

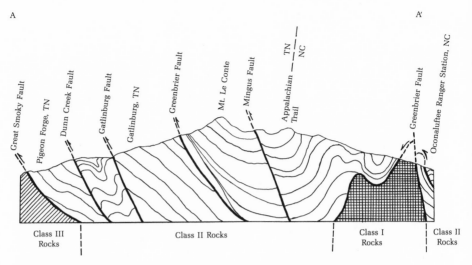

Figure 3. *A geologic cross section of the Great Smoky Mountains National Park illustrates what a "slice" down through the bedrock would look like from the Pigeon Forge, TN, area, across Mt. Le Conte, and finally to the Oconaluftee Ranger Station, NC. The location of this cross section is shown in Map 2 as the line marked A–A'. Note the three classes of rocks which relate to the geologic map.*

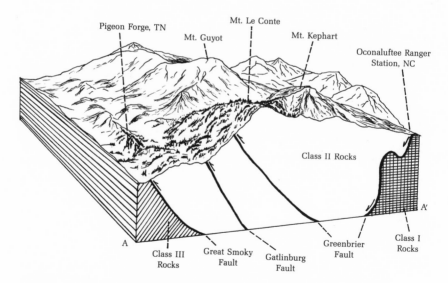

Figure. 4 *This three-dimensional block diagram illustrates the general structure and types of bedrock found in the Great Smoky Mountains National Park. Note that Class II rocks make up the major portion of the Park.*

Table 2. *Rock Types Found in the Great Smoky Mountains National Park*

Class	Formation	Rock Type
III	Lower Ordovician	
	Knox Group	limestone, dolomite
	Lower Cambrian	
	Chilhowee Group	
	Helenmode Formation	siltstone, sandstone
	Murray Shale	siltstone, shale
	Nebo Quartzite	quartzite
	Nichols Shale	siltstone, shale
	Cochran Formation	quartzite, sandstone
	Igneous Intrusive	metadiorite
II	Ocoee Supergroup (Late Precambrian)	
	Unclassified Formations	
	Cades Sandstones and other sandstones	sandstone
	Walden Creek Group	
	Sandsuck Formation	shale
	Wilhite Formation	siltstone, phyllite
	Shields Formation	coarse conglomerate
	Licklog Formation	shale, sandstone
	Great Smoky Group	
	Unnamed Sandstone	sandstone
	Anakeesta Formation	slate, phyllite, schist
	Thunderhead Sandstone	sandstone
	Elkmont Sandstone	sandstone
	Snowbird Group	
	Metcalf Phyllite	phyllite
	Pigeon Siltstone	siltstone
	Roaring Fork Sandstone	sandstone
	Longarm Quartzite	quartzite, arkose
	Wading Branch Formation	siltstone, sandstone
I	Precambrian Basement	schist
		gneiss

Source: P. B. King, R. B. Neuman, and J. B. Hadley, *Geology of the Great Smoky Mountains National Park, Tennessee and North Carolina,* 1968.

Note: The rock classes are those explained in "A View of Geologic History," pp. 31–35.

Look Rock. Strata of the Chilhowee Group and Knox Group constitute the bulk of these rock units.

Fossils

The occurrence of fossils in the Great Smoky Mountains is largely limited to the Cades Cove area and along the Foothills Parkway on Chilhowee Mountain and Green Mountain. Most of the rocks found in the Park are extremely old and bear no evidence of the very primitive life that appeared in the last stages of Precambrian time. However, younger Paleozoic age rocks do contain evidence of past animal life.

The oldest fossils found in the Park are the vertical burrows of a marine animal, believed to be a worm or crustacean. These fossils, known as *Scolithus* sp., appear as closely spaced, narrow cylindrical tubes standing perpendicular to the layers of rock. *Scolithus* can be found in the rocks of the Nebo and Hesse formations of the Chilhowee Group on Chilhowee Mountain. Look for groups of 3- to 8-inch-long vertical lines in the sandstone outcrops along the Foothills Parkway. Fossilized burrows, tracks, and trails are known as trace fossils, impressions left in the sediment by organisms as they reside on, crawl across, or burrow in the earth's surface or ocean floor. The actual animal, however, is not fossilized.

Also found on Chilhowee Mountain are fossils of small shelled animals called ostracods, which repeatedly molted their shells as they grew older. Although these types of crustaceans are largely freshwater, they have occurred in marine waters as well, and it is the bivalved shells of marine ostracods which can be found fossilized in the brown-to-gray shale strata of the Murray Formation at Murray Gap.

The limestone found exposed in Cades Cove contains fossils of thumbnail-sized shelled animals known as brachiopods and fragments of trilobites, scavenging arthropods very similar to present-day horseshoe crabs but only an inch or two in size. Because most layers of the limestone are devoid of the fossils, you may have to look closely to find these traces of early life. Outcrops of gray limestone in the Gregorys Cave area of Cades Cove have also yielded specimens of these fossils (Neuman and Nelson 1965).

Faults

There are a number of faults in the Park, of which four—the Great Smoky, the Gatlinburg, the Greenbrier, and the Oconaluftee—play a major role in the structural arrangement of the bedrock. Moreover, all of these faults have their exposures on the northern and eastern sides of the Park. (The Greenbrier fault is also found in the southeastern section of the Smokies in the vicinity of Balsam High Top Mountain.)

The overall structural configuration of the Smokies may be interpreted as a large syncline with its longitudinal direction (axis) in an east-northeast to west-southwest direction. In addition, the rocks have been contorted and finally faulted and displaced to form a somewhat disjunct group of rock strata.

Cenozoic Deposits and Landforms

Terrestial weathering and erosion, dominant forces during the Cenozoic Era in the Great Smoky Mountains region, have produced an assortment of unconsolidated residual materials and transported deposits. The Cenozoic Era, subdivided into an older Tertiary Period and a younger Quaternary Period, was the time during which most of today's landscapes were formed. Most of the weathering to residual soils, as well as the sculpturing of the mountains and the cutting of the gorges, occurred during Cenozoic time.

During the preceding Mesozoic Era and the Cenozoic Tertiary Period, much of the degradation of the Great Smoky Mountain landmass took place. It is widely accepted that

As streams erode the surface, they also deposit fragments of eroded bedrock in their course. These fragments, ranging in size from sand to large boulders, become rounded because of abrasive encounters with each other and with the stream bottom.

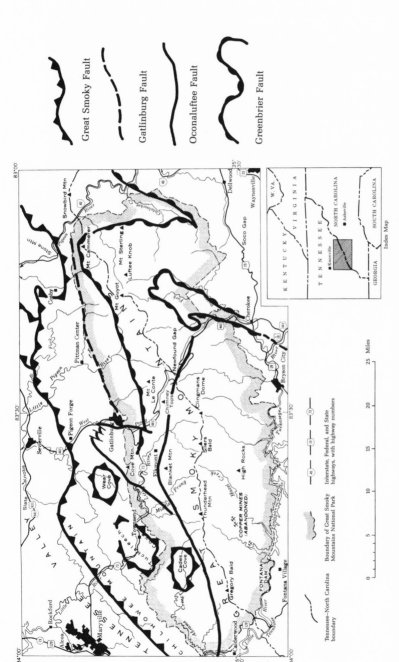

Map 3. *The location of the major faults found in the Great Smoky Mountains National Park. The numerous other faults are of limited extent and therefore are not shown here. Base map after P. B. King, 1949—on back of U.S.G.S. map, "The Great Smoky Mountains National Park and Vicinity" (revised 1972).*

Note: Several shapes of barbs are employed in order to distinguish among the various faults. Each set of barbs, arranged to indicate the direction of a given thrust fault under the land surface, thus also gives a sense of direction of movement and an indication of which rocks have been thrust over the underlying rocks.

Along some sections of the crest of the Smokies, bedrock is exposed, forming jagged ridges. The Sawteeth (near Charlies Bunion) are strata of the Anakeesta Formation.

many erosional episodes took place during this time interval, resulting in several major highland features. Of these features the most notable include the upland surfaces, the valley floor surfaces, and residual and transported materials, including saprolite, residuum, colluvium, and high-level gravels (King 1964).

A group of highland features is collectively known as the valley floor surface, which extends across the foothills of the mountains. This old valley floor surface, marked by the crests of the foothills, forms a gentle slope toward the present Ridge and Valley province in Tennessee. The view from Maloney Point Overlook along Little River Road (Road Log 4, Log Mile 3.5) provides an excellent opportunity to see this ancient set of topographic remnants, beginning at the base of the steeper slopes of Mt. Le Conte and extending to the Ridge and Valley province.

The much higher upland surface is probably a remnant of

Because of the earth's relentless attack upon itself, massive beds of rock eventually end up as smooth-worn boulders in a mountain stream.

This view of Mt. Le Conte and Anakeesta Ridge from the top of the Chimneys shows the gorge of the West Prong of the Little Pigeon River. The upland surface in the distance is the crest of the mountain.

a former landsurface which has now been eroded. Such remnants may constitute the crests of today's high peaks and summits, although they have been downwasted from their original positions. As erosion continued during the Cenozoic Era, the hilly topography was deeply incised, forming the steep-sided gorges that intervene between the higher peaks of today. An eroded area of this type is the gorge of the West Prong of the Little Pigeon River between Mt. Le Conte and Sugarland Mountain (Road Log 2, Log Mile 1.7 to 12.9).

The Quaternary Period, spanning approximately the last 1.8 million years, represents an alternating series of glacial expansions and contractions. Although the area of the Great Smoky Mountains National Park was itself not glaciated, glaciers to the north of it affected the temperatures and precipitation: longer cool cycles, more rain and snow. The Quaternary Period has produced surface deposits which the Park visitor can easily identify, including block fields, fine-textured colluvium, bouldery alluvium, and terraced deposits (Clark and Torbett 1987; King 1964; King and Stupka 1950).

Along the hiking trail to Ramsay Cascades are numerous block fields, some cloaked in vegetation.

In the Greenbrier section of the Park is one of the largest block field deposits in the Smokies. The boulders range in diameter from one foot to over 20 feet.

Block fields consist of accumulations of large, angular blocks and boulders (each up to 30 feet in diameter) which are sheet-like, gently sloping accumulations along the bases of steeper mountain slopes. You can see a number of these block fields along the Newfound Gap Road near the Chimney Tops (Road Log 2, Log Mile 4.5 to 7.5). They occur on or downslope from quartzite or sandstone rock units.

Less conspicuous than the block fields is the fine-textured colluvium, consisting of small, angular rock fragments, usually in a matrix of silty clay soil. Most often colluvium is locally derived, originating from rock units nearby or underlying the deposit a short distance upslope.

Bouldery alluvium is found in small, gently-sloping coves which are floored with a heterogenous mixture of sandy loam and pebbles with numerous protruding rounded boulders. The boulders probably originated in the block fields high up on the mountain slope and were transported to their present site by erosional processes during the colder (Ice Age) episodes of the Quaternary Period. One such cove floored with bouldery alluvium is on the West Prong of the Little

Block fields which developed during the Pleistocene Epoch can be seen along U.S. 441 near the Chimneys picnic area. Bluffs of Thunderhead Sandstone located upslope are the source of the boulders.

Pigeon River in the Sugarlands (Park Headquarters) area; another is in the Greenbrier section of the Park (Road Log 1, Hiking Trail to Ramsay Cascades).

The stream-derived terraced deposits, usually occurring in the foothills section and in the adjoining Ridge and Valley province, consist of small pebbles, cobbles, and boulders, whose edges were well rounded by the action of streams and rivers in the past. Now, however, these deposits are found on topographic terraces which may be 100 feet or more above the modern stream floodplain. You can see this type of deposit along S.R. 73 near its junction with S.R. 32 at Cosby, Tennessee (Road Log 1, Log Mile 20.6).

Recent studies have resulted in the development of a general model for the late Quaternary history of the Blue Ridge province, including the Great Smoky Mountains National Park (Shafer 1984; Delcourt and Delcourt 1979, 1981, 1985). It is believed that approximately 20,000 to 16,500 years before the present, the mountain peaks of the Great Smoky Mountains region had tundra vegetation and had developed permafrost where the mean annual temperatures were below 32° F.

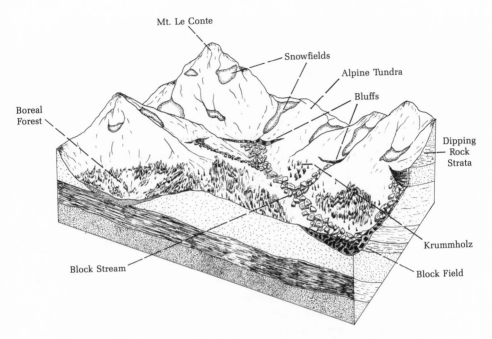

Figure 5. *This schematic block diagram illustrates the environmental conditions prevalent approximately 16,500 to 20,000 years before the present on Mt. Le Conte (as seen from U.S. 441: Road Log 2, Mile 2.1). Adapted from Delcourt and Delcourt, 1985.*

In fact, a permanent snowpack may have persisted throughout the year in some higher hollows. Intense freeze-thaw activity resulted in the development of block fields. Alpine tundra herbs and subarctic shrubs persisted above 4,950 feet in elevation. Boreal forests blanketed the hill slopes and valleys at lower elevations. The upper limit of stunted trees (called *Krummholz*) occurred at an altitude of about 4,950 feet.

During the time between 16,500 and 12,500 years before the present, there was an increase in mean annual temperature and precipitation. Mass wasting and freeze-thaw action reworked sediments down the unstable mountain slopes. With warming climates, boreal forests spread into the middle elevations and deciduous tree species (such as oak, birch, and ash) migrated into the valley, expanding from refuge areas in the coastal plain.

Major block field movements and colluviation took place

between 12,500 and 10,000 years ago, when the mean annual temperature was around 32° F and precipitation values were high. During this time interval, colluvial processes gave way to stream erosion of the upland surfaces. By 10,000 years ago boreal coniferous forests—dominated, as today, by Fraser fir and red spruce—were established on the higher peaks; oak forests spread into the low and middle elevations.

Today spruce-fir forests are found only at the crests of the highest peaks. Relict populations of boreal and subarctic plant species are found along cliff faces on Mt. Le Conte. Block fields, now stabilized, support a growth of hemlock and hardwood trees.

Although it was once thought that the rich forest communities found on colluvial and block field surfaces in the Smokies were relicts of the Tertiary Period, recent studies suggest that these mature forests with large trees can date only from the Holocene Epoch, the last 10,000 years of modern warm climate (Delcourt and Delcourt 1985). During most of the Quaternary Period, the landscape was in an unstable state, with the continued movement of boulders and active colluvium preventing the establishment of hardwood vegetation. Only after the landscape had stabilized, by 10,000 years ago, did the hardwood trees spread into the Great Smoky Mountains. Today you can observe these mature forests on block fields and colluvium at several locations in the Park, including the Locust Nature Trail and the Buckeye Nature Trail in the Chimney Tops area and along the foot trail to Ramsay Cascades in the Greenbrier section.

Along the high ridges and mountaintops of the Smokies are vegetational anomalies known as balds. Two types of balds can be distinguished in the Smokies: heath balds and grass balds. Found on very rocky slopes and ridges, the heath balds are covered by plants of the heath family (rhododendron, mountain laurel, etc.). These thick, entangled masses of shrubs, appearing smooth or slick from a distance, can be seen along U.S. 441 from near the Chimney Tops to past Newfound Gap into North Carolina (Road Log 2).

The grass balds are found primarily along the crest of the Smokies from near Clingmans Dome westward to the Park boundary near Fontana Dam. Some speculation continues about their origin, but recent studies indicate that the grass balds are disappearing as the surrounding hardwood forests encroach upon them. There appears, however, to be no relationship between the bedrock and the balds (King 1964). Although the origin of the grass balds is unknown, they seem to have been maintained artificially because the early settlers used them as summer pastures for their livestock. Whether these balds were formed partially by fire or some other natural phenomenon, their existence has been documented in Native American traditions for at least two hundred years.

Summary of Geologic Events

Precambrian Time. Early in the geologic history of the Great Smoky Mountains, the formation of the Precambrian basement complex involved the metamorphism of accumulated marine sediments and igneous rocks. Radioactive dating methods indicate that the metamorphism and deformation of the basement complex took place approximately one billion years ago. An erosional period characterized the top of the basement complex prior to the formation of younger Precambrian rocks.

In late Precambrian time the deposition of mud, silt, and sand in an ocean resulted in the formation of the rocks that make up the Ocoee Supergroup. These sediments accumulated on a broad continental shelf adjacent to an eroding landmass. Approximately 50,000 feet of sediments were deposited as the ocean encroached on the landmass. Most of the rock formations found in the Park are classified as being within the Ocoee Supergroup. Fossils have not been found in these rocks.

Paleozoic Time. Deposition of marine sediments was characteristic of the time period from the Cambrian to the Mississippian (300 to 600 million years ago). In an area which lies adjacent to but north and west of the Park, marine sediments were being deposited on a slowly submerging continental shelf. Shale, siltstone, and sandstone compose most of the Cambrian age rocks (Chilhowee Group). Beginning late in the Cambrian period, carbonate rocks (limestones) became increasingly abundant until the close of the Mississippian pe-

riod. By the end of the Paleozoic Era, many thousands of
feet of sediment had been deposited, forming the many rock
units of the Ridge and Valley province, which now stretches
from northern Georgia and Alabama to Pennsylvania and
New York.

By the middle of Ordovician time, the continental drift
mechanisms had been set in motion. Plate tectonic theory,
which includes continental drift, postulates that as the crustal plates collided, the proto-Atlantic Ocean was destroyed
and the supercontinent of Pangea was produced. The con-
vergence of continental plates provided the forces to deform
the rocks of the Appalachian Mountains.

During early Paleozoic time (after the rocks of the Ocoee
Supergroup were initially deformed), igneous intrusions con-
sisting of dikes and sills developed in the Ocoee Supergroup.
The igneous activity was the result of the early stages of the
Appalachian mountain-building episode.

The Greenbrier fault is believed to have formed during the
early Paleozoic Era as a result of tectonic activity in the northern Appalachian Mountains (Taconic Orogeny). Subsequent
periods of erosion and deposition culminated in a final
mountain-building episode of the Paleozoic Era known as the
Appalachian or Allegheny orogeny. As a result of the tectonic
activity, the Greenbrier fault has undergone multiple episodes of deformation, distorting the fault plane itself.

The rocks of the Ocoee Supergroup and the Cambrian age
formations were metamorphosed, folded, and faulted. The
displacement of a very large block of crustal rock was produced by a series of plate movements, pushing the Basement
Complex, Ocoee Supergroup rocks, and older Cambrian strata
to the northwest. The result was the formation of the Great
Smoky fault (and its associated faults). It is believed that near
the close of Paleozoic time the Gatlinburg fault was emplaced.

Mesozoic Time. The Great Smoky Mountains were being
shaped by erosional processes during the Mesozoic Era. Much
of the younger rock strata were eroded away, leaving only the
much older rock exposed. Because the continental landmass
continued to experience uplift at the same time it was being
eroded, the series of ridges, valleys, and mountain crests
developed.

There is evidence that the mountain rocks, thrust over the eastern half of the Valley and Ridge province by the Great Smoky fault, were eroded away during the Mesozoic Era. The result was that the eastern portions of the present-day Valley and Ridge province were uncovered by erosion of the faulted overlying mountain rocks (much like the erosional development of the Cades Cove and Townsend areas today).

As the Great Smoky Mountains were being carved out of the bedrock, a new ocean basin began developing to the east and south. The continental plates of Europe and Africa began their drifting journey away from North America and formed the present-day Atlantic Ocean. The climate of the Great Smoky Mountains region changed from tropical to subtropical and then to temperate as the North American continent moved to more northerly latitudes.

Cenozoic Time. Terrestrial erosion continued to carve and shape the bedrock surface during the Cenozoic Era. Repeated uplift and erosion resulted in the development of regional drainage patterns in the Great Smoky Mountains area. During the cold phases of the Pleistocene Epoch (from 10,000 to 1.8 million years before the present), periglacial processes resulted in the formation of numerous surface deposits and landforms, such as block fields, fine-textured colluvium, alluvial deposits, and sculptured peaks, slopes, and valleys. Within the limestone coves and across adjacent portions of the Ridge and Valley province underlain by limestone, numerous solution caves were formed. These caves provided shelter for a number of now extinct mammals, including the saber-toothed tiger and the giant sloth. Other giant mammals such as the wooly mammoth and the mastodon were also present in the low-lying areas of the region.

Even today shaping of the Great Smoky Mountains landscape continues; the dominant erosional force is now fluvial activity. Although we can track the effects of some processes like erosion, others—notably continental drift—are too slow for us to notice. To us the mountains appear mostly static, with a stable climate. In the warm, humid climate that now exists, the mountain slopes abound with flora and fauna: about 2,000 species of plants, 50 species of mammals, 39 species of amphibians, and over 200 species of birds.

Rock slides occasionally close U.S. 441, which crosses the crest of the Smokies. The slides usually involve bedrock, soil, and vegetation which cascade down the mountain slope very rapidly in a jumbled mass.

Natural Geologic Catastrophes

Of the events that have shaped the history of the Great Smoky Mountains, geologic catastrophes—flash floods, debris flows, avalanches, and landslides—have had the most recent impact.

Because the Smokies are located in the temperate climate of southeastern North America, the Park is subject to very heavy rainfall in short periods of time (usually during the spring and summer). As much as 8 to 12 inches of rain may fall in a 24- to 36-hour period, swelling creeks and rivers well beyond their carrying capacity. The result is usually a wall of water, often accompanied by trees, mud, rocks, and other debris, which flows down the stream bed. Occasionally these debris flows lodge and dam the stream, forming small ponds or lakes. Although the debris flows and avalanches are capable of changing the course of a stream, their effects are usually local in nature. (Hikers, however, must be aware of this everpresent danger and alert to local weather forecasts.)

Another type of catastrophic event that happens in the Smokies are landslides, usually a specific sort known as wedge failures. Wedge failures occur as a result of the intersection of the two planes of weakness in the bedrock that have developed as a result of the folding and faulting and crushing of the bedrock during its geologic past. The movement oc-

Peaceful mountain streams can become torrents of rushing water when summer thunderstorms produce flash floods.

Debris slides occurring as a result of a wedge failure landslide leave massive scars on the landscape, as here on Anakeesta Ridge near New-found Gap.

This "V"-shaped scar near Anakeesta Ridge is a result of a wedge failure landslide.

The energy produced by a rain-swollen stream can move large boulders along its course.

During debris slides the moving rock material acts as a grinding machine, reducing trees and shrubs to gnarled pieces of wood.

A landslide scar on Anakeesta Ridge resulted from a wedge failure and debris slide after heavy rains. The rock strata involved in the slide belong to the Anakeesta Formation.

In the spring of 1986, a wedge failure landslide produced its typical "V"-shaped trough along the base of Anakeesta Ridge. U.S. 441 is visible in the background.

curs along the intersection of the two planes of weakness. As a result of the slide, a wedge-shaped scar is left on the landscape and usually takes many years to heal.

These types of landslides have occurred naturally on Mt. Le Conte and Anakeesta Ridge (Clark, Ryan, and Drumm 1987). The slide scars are very noticeable on U.S. 441 about 2 to 3 miles down from Newfound Gap on the Tennessee side (Road Log 2, Log Mile 11.2). Many of the very narrow "V"-shaped ravines located in the Smokies are probably the result of natural wedge failure landslides.

In the summer of 1984, two wedge failures occurred simultaneously on U.S. 441 about 4 miles from Newfound Gap on the Tennessee side of the Park. As a result, the highway was completely blocked, stranding over 20 cars between the slides. The smaller of the two was cleared within a few hours, but the larger required several weeks of work.

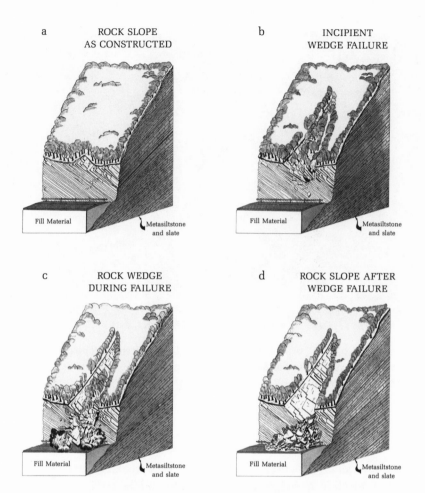

Figure 6. *A wedge-failure landslide (Moore, 1986).*
a. *A natural rock slope, as on Anakeesta Ridge in the central part of the Smokies, or a man-made roadway cutslope, as illustrated above, may have the potential for wedge-failure landslides.*
b. *The incipient development of a wedge failure may include cracks opening in the ground and trees leaning or tilting in response to ground movement. Serious rockfall may also be an indicator.*
c. *A wedge-failure landslide during maximum movement results in a very powerful force which destroys everything in the path of the movement. During failure the rock mass moves very rapidly, usually in a matter of seconds, covering and removing almost everything downslope in its path.*
d. *The results of a wedge-failure landslide are a large wedge-shaped scar (exposing bedrock) left on the landscape and a large mound of rubble and debris left at the bottom of the slope.*

The block diagrams (Figure 6) illustrate the development of a wedge failure and its resulting scar. Figure 6a shows a normal rock slope that results from a roadway cutslope. (A natural slope could have these same conditions.)

The initial or incipient development of the wedge failure is illustrated in Figure 6b. Small cracks, developing high on the slope, provide access for additional water to infiltrate and lubricate the slide mass. In addition, trees may begin to show signs of slumping or bending as a result of the moving earth beneath them.

As shown in Figure 6c, the mass movement of the wedge material is usually very violent and destructive. Trees are crumpled, rock is broken, and anything downslope in its path will be covered or obliterated.

The final stage of this catastrophe results in a wedge-shaped scar (Figure 6d): bare rock surfaces devoid of vegetation extend up or down the mountainside for several hundred yards. A mass of debris and rock rubble is mounded at the "toe" or bottom of the landslide.

These wedge failures usually occur very quickly—within a few minutes—without warning. Luckily no one has been injured in one of these landslides, although some wildlife probably have fallen prey to these destructive forces. So as you enjoy the beauty and splendor of the Smokies, remember that geologic catastrophes still occur and can be very violent. (The road logs note where these landslides have taken place and are likely to reoccur.)

A Note on Geologic Mapping

When geologists want a better understanding of the geology of an area, one of the first things they do is make a geologic map of the surface exposures of bedrock. Information put on the map may include the rock type (e.g. limestone, granite, shale), structural character of the beds of rock (i.e. horizontal or dipping, and in what direction), and the extent of the rock exposure. In addition, other features such as faults, folds, and mineralization may be noted on the map.

Most of the geologic mapping in the Great Smoky Mountains National Park was completed in the period from 1946 to 1956 (King 1964). The earliest known work to include significant observations on the area was by the Tennessee geologists James E. Safford in 1856 and 1869 and Arthur Keith in 1889 and 1894 (King 1964).

Trying to locate all of the surface exposures of bedrock in the Smokies is no trivial task. Discerning the configuration of the bedrock is made even more difficult by the thick, lush vegetation and the deeply weathered land surfaces. Geologists working in the Park must climb the ridges and follow the stream beds, searching for any rock exposure that may give a clue to the contents of the subsurface. In earlier days this was done on horseback and on foot. Even today long hikes into the Park are necessary to acquire detailed geologic information. However, the modern technology of roadway cutslopes and satellite imagery makes matters easier for today's geologists.

Because of the vegetation the Park visitor may never see some of the bedrock exposures or fault contacts which are described in the road logs. In most cases the forest cover obscures the bedrock, with only a change in the topography hinting at a possible fault or change in bedrock. Abrupt changes in the overall landscape are visual clues to a change in the subsurface. The most obvious of these clues is the separation of the foothills of the Smokies from the higher slopes and main crest by the Gatlinburg thrust fault (Figures 3 and 4).

A very useful geologic map of the Smokies is included in *Geology of the Great Smoky Mountains National Park, Tennessee and North Carolina* (U.S. Geological Survey Professional Paper 587) by Philip King, Robert Neuman, and Jarvis Hadley. Originally published in 1968, this short paper (23 pp.) has been reprinted and can be purchased at the Sugarlands or Oconaluftee Visitor Center in the Park.

PART 2

**Road Logs
and
Hiking Trails**

Using the Road Logs

This section of the book provides geologic road guides to five well-known, scenically spectacular—and geologically significant—highways in the Park or along its boundary. Each road log is accompanied by a log for a hiking trail that intersects with it.

Because these logs are meant to be pointers in the field, as you travel, they are not encumbered with detailed geologic analysis. In the earlier expository sections of the book and in the glossary, you can find more extensive explanations of the terms used and formations described.

Each log begins with a brief description of the route, followed by data about the distances and altitudes traversed and the general character of the road. Accompanying each road log is a brief list of relevant terms with which you might wish to reacquaint yourself before setting out. Strip maps showing cultural features appear with each log. In addition to photographs of particular structures and formations, each log includes cross sections and three-dimensional block diagrams of the overall geology along the road.

Each log begins at a clearly designated point with a Log Mile of 0.0 (usually at the Sugarlands Visitor Center at the Park Headquarters in Gatlinburg) and continues to the end of the route. The notable rock exposures and other points of interest you can see are listed along with your distance from the beginning. At some point along each road log there is the starting point for a convenient and interesting hiking trail. Each hiking trail is presented separately, immediately following the appropriate road log. Distances listed in the headnotes for both road logs and hiking trails are one way.

In most instances where the road trip crosses a major fault or a boundary between two rock formations, you will not be able to see the actual fault or the contact between the two formations. However, you will notice a gradual change in the scenery—a change in topography, vegetation, and rock exposures.

When references to rock outcrops are made, these rock exposures are generally within easy sight of the road (or hiking trail). Scenic vistas will usually make you want to get out your binoculars.

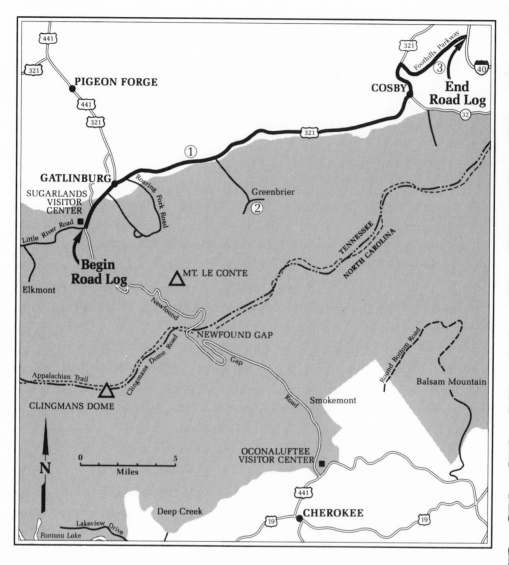

① Greenbrier Pinnacle vista, Log Mile 7.7

② Beginning of Hiking Trail to Ramsay Cascades

③ Green Mountain vista, Log Mile 24.8

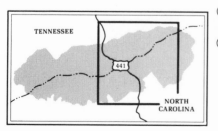

Road Log 1

Route: S.R. 73 from Great Smoky Mountains National Park
Headquarters to I-40 at Cosby, Tennessee (via Foothills
Parkway)
Beginning elevation: 1,475 feet
Ending elevation: 1,320 feet
Maximum elevation attained: 2,250 feet
Trip length: 28.4 miles
Nature of the road: Two-lane, paved highway, with numer-
ous straight sections of road and some curvy portions
Special features: Scenic view of Greenbrier Pinnacle and
Mt. Cammerer (east end of Park); connection with I-40
Hiking trail (optional): Ramsay Cascades
For reference: Gatlinburg fault, colluvium and block fields,
terraced deposit

On this road trip you will travel along the north side of the
Park boundary, through Gatlinburg and Cosby, crossing
Green Mountain via the Foothills Parkway, and terminating
at Interstate 40. This area is underlain by complexly folded
and faulted rocks that are Precambrian in age. These rocks
are sedimentary in origin (derived from ocean sediments)
and consist mainly of sandstone and siltstone, whose names

refer to the size of the grains in the rock (i.e., sand-sized or silt-sized). Geologists term these particular underlying rocks the Ocoee Supergroup. Principal formations in the area include Pigeon Siltstone, Roaring Fork Sandstone, and Thunderhead Sandstone. Small portions of a number of other formations crossed are noted in the log.

Traces of major faults—including the Gatlinburg fault, the Great Smoky fault, and the Dunn Creek fault—will be crossed during the trip. For the first twenty miles, the trace of the Gatlinburg fault is adjacent to and paralleled by the roadway you will be traveling on.

Important features to watch for include roadway cutslopes exposing bedrock in the Gatlinburg area, spectacular views of the crest of the Smokies—including the summits of Mt. Le Conte, Greenbrier Pinnacle, Mt. Guyot, Cosby Knob, and Mt. Cammerer—and rock exposures and vistas along the Foothills Parkway.

Road Log 1

Mileage	Description
Mileage	*Description*
0.0	Leave Sugarlands Visitor Center at the Park Headquarters (intersection of U.S. 441 and Little River Road). Turn left onto U.S. 441 North.
1.7	On leaving the Great Smoky Mountains National Park, notice the dark-gray moss-covered rocks on the hill to your right: these are exposures of Roaring Fork Sandstone. Next you travel through the town of Gatlinburg, which is built mainly on boulders and sand; these colluvial and alluvial deposits of the West Prong of the Little River and Le Conte Creek washed down from the higher slopes of the Smokies. Once known as Whiteoak Flats, the Gatlinburg area is a small cove in the foothills of the Smokies.
3.2	Turn right onto S.R. 73 East.
4.9	On your left you will see flat planes of grayish-brown rock dipping toward the road. These are scars of a man-caused landslide which has exposed rock units of Roaring Fork Sandstone. This rock slide was originally caused by exposing the dip plane of the bed-

ding and fracturing the rock during roadway construction. Recent road improvements, such as the small retaining wall at your left in the curve, have stabilized landslide movement. For approximately the next 18 miles, S.R. 73 will parallel the Gatlinburg fault. Although geologists know that it is located along the right side (south) of the road, the actual fault is not visible; not even a change in the vegetation or land surface can be seen. What you can see, however, are rock formations associated with this fault: strata of Roaring Fork Sandstone which have been thrust upon younger strata of Pigeon Siltstone. Exposures of Pigeon Siltstone and the soils derived from its weathered residue can be observed in roadway cutslopes along the road up to Log Mile 20.8. The Pigeon Siltstone formation consists of thin beds of tabular or slabby rock, greenish-gray to grayish-brown; the residual soils are reddish-orange to yellowish-brown, with small chips of shaly rock mixed in.

7.7   Dramatic view of Greenbrier Pinnacle (approximate elevation: 4,500 feet). Greenbrier Pinnacle is composed of strata of Thunderhead Sandstone. This very durable, weather-resistant sandstone is what you see if you use your binoculars to look at the 150-foot-high bluff near the top of the pinnacle. Approximately halfway down the slope of the pinnacle, Thunderhead Sandstone is thrust-faulted upon Roaring Fork Sandstone, forming a cap of very resistant rock strata. These same strata extend up to the top of Mt. Guyot, the high peak directly behind Greenbrier Pinnacle at the crest of the Smokies (Figure 7). Behind the pinnacle but not visible from the road is a beautiful waterfall called Ramsay Cascades. Massive cliffs of Thunderhead Sandstone form the prominence over which a mountain stream flows, forming the cascade. A hike to Ramsay Cascades makes an excellent side trip for a close view of how the massive Thunderhead Formation occurs in the park (see Hiking Trail 1, pp. 79–83). There is also a trail to the top of Greenbrier Pinnacle.

Most of the bedrock during the first 18 miles of this road trip is underlain by metasiltstone and metasandstone strata. The bedrock is badly fractured as a result of nearby thrust faults.

At Log Mile 7.7 is a scenic view of Greenbrier Pinnacle. Massive bluffs of outcropping Thunderhead Sandstone are near the top of the pinnacle.

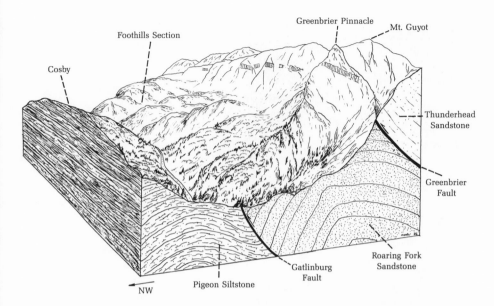

Figure 7. *The geologic conditions around Greenbrier Pinnacle are illustrated by this block diagram. This view is from S.R. 73 at Log Mile 7.7. Note how the Thunderhead Sandstone brought to the surface by the Greenbrier thrust fault dominates the crest of the Smokies.*

9.0   Greenbrier entrance to the Great Smoky Mountains National Park. If you wish to hike to either Ramsay Cascades or Greenbrier Pinnacle, you should turn to enter here. At this point, continuing straight on S.R. 73, you also cross the Little Pigeon River as it leaves the Park, draining the Greenbrier section. Take a look at the boulders in the stream bed: the flat, tabular rocks consist of slate and phyllite; the rounded gray to milk-white ones are quartz; sandstone and metasandstone are also present. All are eroded fragments of Roaring Fork Sandstone, Elkmont Sandstone, and the Anakeesta Formation.

11.8   Cross Webb Creek.

17.6   Baxter's Apple Orchards. Producing mainly Winesap apples, these orchards are developed on a mantle of residual silty clay soils of Pigeon Siltstone strata.

Tilted strata of Roaring Fork Sandstone along the entrance road to the Greenbrier section of the park (Log Mile 9.0). The rock strata dip south in this view of the Little Pigeon River.

Greenbrier Pinnacle, as seen from S.R. 73.

From a different angle Greenbrier Pinnacle looks like a moderately sloping ridge instead of a dominating pinnacle. (View is from S.R. 73 near Log Mile 12.)

Mt. Cammerer (4,928 feet) dominates the skyline of the Smokies near Log Mile 20.

17.8   Sevier-Cocke County Line.

20.6   On your right, in the excavated area, you can see gray to light brown rounded gravels and cobbles in sandy soil. These are ancient alluvial deposits, perhaps of nearby Cosby Creek at a time when the existing valleys were much higher in elevation. It is thought that these terraced deposits were probably formed during Late Pleistocene time, the last Ice Age epoch in North America. They were produced by stream erosion and deposition when large volumes of material were available in the mountains as a result of mass wasting under rigorous climatic conditions, and when stream run-off was at a maximum (King 1964).

20.8   Turn left on S.R. 32; enter Cosby community.

21.5   Here you cross the Dunn Creek fault, which is exposed in the road cut along the sharp curve to your left. The brownish-white to buff-brown sandstone/siltstone exposed on the left of the cut has been

An old terraced alluvial deposit—perhaps an ancient location of nearby Cosby Creek—is exposed in a road-cut and adjacent excavation near Log Mile 20.6 on S.R. 73 (about 0.2 mile before the S.R. 32 intersection at Cosby, TN).

Green Mountain (viewed from S.R. 73 in Cosby, TN) is underlain by Cambrian age rocks of the Chilhowee Group. The route of the Foothills Parkway can be seen above.

Near Log Mile 21.5 a roadcut on S.R. 32 exposes the Dunn Creek fault. This view clearly shows the fault line: Pigeon Siltstone appears in the left two-thirds of the picture and the shaly Wilhite Formation in the right third.

thrust upon the reddish-brown shaly strata of the Wilhite Formation. The Dunn Creek fault is located where the shaly rock comes in contact with the more blocky sandstone.

21.8 On your right in the river gorge, the outcrops of gray-jagged rock are of the Chilhowee Group (Class III type rocks: Table 2). Activity of the Great Smoky fault, itself not visible, has thrust older strata of the Wilhite Formation (Class II) onto the younger Chilhowee Group. The Wilhite Formation, composed of a shaly rock which weathers more rapidly than sandstone, forms part of the valley floor in this area but cannot be seen in the river gorge.

22.0 Cross Cosby Creek.

22.6 Turn right off S.R. 32 onto the Foothills Parkway. This portion of the road trip ascends Green Mountain, which is underlain by a quartz sandstone belonging to the Chilhowee Group. On this section of the Foothills Parkway you get excellent views of the north flank of the Great Smoky Mountains National Park and Cosby Valley below. The concrete retaining wall on your left was constructed to restrain a landslide which was precipitated when the roadway was cut through old colluvial and residual material of the Chilhowee Group.

24.8 Be sure to stop at the scenic overlook to the right (this overlook is near the top of the mountain; you will already have passed a small overlook approximately halfway up the mountain). Here you will have fine views of the Cosby Valley below, Mt. Cammerer, Cosby Knob along the crest of the Smokies.

The cutslope behind the overlook is composed of Hesse Quartzite of the Chilhowee Group. Notice not only the hard, dense bedrock but also the numerous joints (fractures) in the rock strata which resulted from the nearby thrust-faulting activity of the Great Smoky fault at the base of Green Mountain. The Park Service has installed a short test strip section of black wire meshing on the sandstone cutslope as an experiment to see how well the wire mesh would hold back loose rock and also to promote growth of vegetation.

As you look across the Cosby Valley, you will see two very prominent mountain peaks, Mt. Cammerer to the left and Cosby Knob to the right. The crest of Mt. Cammerer slopes down eastward (to the left from the overlook) to Davenport Gap, the eastern terminus of the Great Smoky Mountains National Park.

The mountain crest from Mt. Cammerer to Cosby Knob is underlain by strata of Thunderhead Sandstone. The Greenbrier fault (not visible) has thrust strata of Thunderhead Sandstone onto strata of Rich Butt Sandstone; the fault plane dips south (back into North Carolina).

The accompanying block diagram (Figure 8) illustrates the sequence of rocks underlying the mountain and valley at this stop. Cosby Valley is dissected by three additional thrust faults: in order, beginning at the base of Green Mountain (the mountain that you are on now) and proceeding toward the main crest of the Smokies, they are the Great Smoky fault, the Dunn Creek fault, and the Gatlinburg fault. The Great Smoky fault has thrust strata of the Wilhite Formation onto the younger Chilhowee Group. The Dunn Creek fault has thrust strata of the older Pigeon Siltstone onto the younger Wilhite Formation. The Gatlinburg fault has thrust strata of Roaring Fork Sandstone onto the younger Pigeon Siltstone. These rock units and faults are not visible from the overlook but are hidden beneath the valley floor, covered by clay soils derived from weathering of the rocks.

The rock units in this complex array of thrust faults have been so broken up that they have been more subject to weathering than the surrounding mountains. As the surfaces eroded, the Cosby Valley was formed. The valley floor is composed of rich and fertile silty clay soils forming a mantle over the broken bedrock. This fertile soil has enabled the

The Thunderhead Formation makes up the crest of the Smokies in the Cosby section of the Park. Mt. Cammerer is the dominant peak in this scenic view from Green Mountain at Log Mile 24.8.

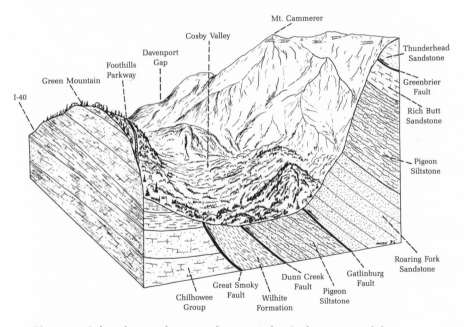

Figure 8. Subsurface geologic conditions in the Cosby section of the Park. Note that the crest of the Smokies is capped by Thunderhead Sandstone. The scenic overlook on Green Mountain along the Foothills Parkway is located along the left side of the diagram.

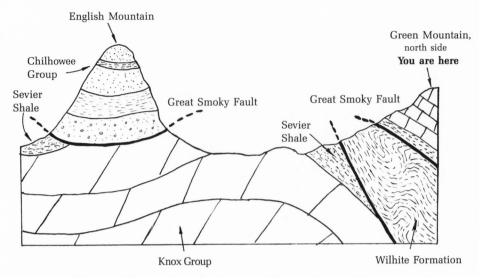

Figure 9. *The scenic overlook at Log Mile 26.2 faces northwest toward English Mountain in the distance. The Great Smoky thrust fault is found at the base of English Mountain and the base of Green Mountain but has been eroded away between the two.*

At the entrance to Interstate 40 from the Foothills Parkway (end of road log), Mt. Cammerer is seen rising almost 4,000 feet above the Pigeon River Gorge.

Cosby Valley to become the home of numerous farms and fruit tree orchards.

25.1   Boundary of the Cherokee National Forest.

26.2   If you stop at the scenic overlook on the left, you will have a fine view to the north across a portion of the adjacent Ridge and Valley province and of English Mountain (far away and to the left). The valley floor between this overlook on Green Mountain and English Mountain in the distance is made up of limestone and shale, which weather and erode more quickly than the rocks beneath the two mountains. The result is the rolling terrain of the valley before you.

English Mountain is composed of strata of the Chilhowee Group which have been thrust-faulted over younger limestone strata of the Knox Group. The bedrock of the mountain—consisting of very hard and weather-resistant quartzites, sandstones, and shales—has been folded into a downward warp known as a syncline. Accordingly, English Mountain is known as a synclinal mountain. The accompanying cross section illustrates these structural aspects.

26.7   Leave boundary of Cherokee National Forest.

27.5   Small overlook to the left.

28.4   Enter interchange with I-40.

End Road Log 1

The Ramsay Cascades are formed where Ramsay Prong flows over a bluff composed of Thunderhead Sandstone. The massive sandstone strata dip away from the viewer at about a 30° angle.

Hiking Trail 1: Ramsay Cascades

Location: At end of Ramsay Prong road in Greenbrier sec-
tion of Great Smoky Mountains National Park, Tennessee
U.S.G.S. 7½ minute quadrangle map: Mt. Guyot, N.C.
Start: End of Ramsay Prong road, 2,080 feet
End: Ramsay Cascades, 4,300 feet
Nature of trail: Unpaved foot trail; maximum elevation gain,
2,220 feet
Distance: 4.0 miles

The Ramsay Cascades trail provides for a scenic and rewarding hike along Ramsay Prong, which flows through a mature
forest of hardwoods and hemlock in the Greenbrier section
of the Park. Two major geologic features are found along this
trail: Pleistocene (Ice Age) colluvial and alluvial deposits and
exposures of Thunderhead Sandstone. (Colluvium consists of
boulders deposited by gravity; alluvium is stream sediment.)
    Both the road leading from S.R. 73 to the trailhead and
most of the Ramsay Cascades trail pass over extensive deposits
of colluvial block fields (gray, angular boulders), fine-textured
colluvium (soil), bouldery alluvium, and recent alluvium. At
the end of the trail, massive outcrops of Thunderhead Sand-

As the climate warmed approximately 10,000 years ago, block fields that accumulated in the ravines became the boulder-filled Ramsay Prong and the Middle Prong of the Little Pigeon River. This view of the upper portion of the latter illustrates how water can transform angular boulders of colluvium to rounded boulders of a stream.

stone form the precipice over which Ramsay Prong flows, creating the sixty-foot-high cascade.

The trail also passes through a virgin stand of cove hardwoods and hemlocks. Some of the largest trees in the Park are found along this trail: tulip poplars (5-foot diameter), sweet birch (3.5-foot diameter), silverbell (2.5-foot diameter), and black cherry (3-foot diameter). In the spring this is a particularly beautiful trail for wildflowers.

Trailhead. To reach the beginning of the Ramsay Cascades trail, turn off S.R. 73 at the Greenbrier entrance to the Park (Road Log 1, Log Mile 9.0) just east of Gatlinburg, Tennessee. Proceed along the Park road for approximately 3.1 miles, until the road splits. Turn left, go across the bridge, and follow

The trail to Ramsay Cascades passes across numerous block fields of angular sandstone boulders originating from massive cliffs of Thunderhead Sandstone located upslope.

the road to the parking area (approximately 4.7 miles from S.R. 73).

Trail details. The Ramsay Cascades trail follows an old roadbed for about 1.5 miles along the Middle Prong of the Little Pigeon River. This roadbed passes through both the gray, angular boulders of numerous block fields and old, bouldery alluvium, stream-worn and rounded. Some of the block fields, containing large sandstone boulders thirty to forty feet in diameter, stretch several hundred yards up the mountain slopes. It is thought that these block fields were produced approximately 10,000 to 12,000 years ago.

At 1.5 miles the graveled roadbed ends and the foot trail to Ramsay Cascades continues. (The trail to Greenbrier Pinnacle turns to the left near the end of the road; the Ramsay Cascades trail proceeds along the creek at the very end of

Some of the Park's largest tulip poplars and hemlocks can be seen along the trail to Ramsay Cascades. The large trees have grown among boulders of block fields which have now stabilized.

Massive outcrops of Thunderhead Sandstone along Ramsay Prong in the Greenbrier section of the Park.

the road.) Where the trail crosses Ramsay Prong at 2.2 miles, a massive outcrop of dark-gray, moss-covered Thunderhead Sandstone can be seen across the creek, downstream from the footbridge.

At about 2.5 miles the trail passes through a stand of very large tulip poplars and hemlocks. Some of the trees, up to five feet in diameter, are estimated to be over three hundred years old. This mature forest developed on a Pleistocene block field whose ruggedness was tempered by 10,000 years of weathering and vegetative growth. Most of the boulders here are hidden by a mat of vegetation.

Near 3.0 miles the trail crosses a footbridge over Ramsay Prong and traverses numerous boulder deposits. These boulders, both colluvial and alluvial in origin, vary in size and shape. The trail ascends a steep slope and rejoins the main channel of Ramsay Prong at 3.5 miles. For the last half-mile the trail twists and climbs through a block field of moss-covered colluvium.

At the base of the sixty-foot-high Ramsay Cascades, the trail ends. Massive exposures of Thunderhead Sandstone form a precipitous bluff over which Ramsay Prong flows. You can see that the sandstone dips away from you at about a thirty-degree angle. Massive exposures of sandstone, such as these at the cascades, were sources of the block fields and other boulder deposits produced during the Pleistocene Age. Until about 10,000 years ago, the area was free of forest vegetation and the boulders could easily move down the mountainside, propelled only by gravity.

Caution: Climbing on the rocks around and above the falls can prove very hazardous and therefore is not recommended. During the winter months ice buildup around the cascades sometimes produces a beautiful ice-blue natural sculpture; however, even the approach to the falls can be dangerous in the winter.

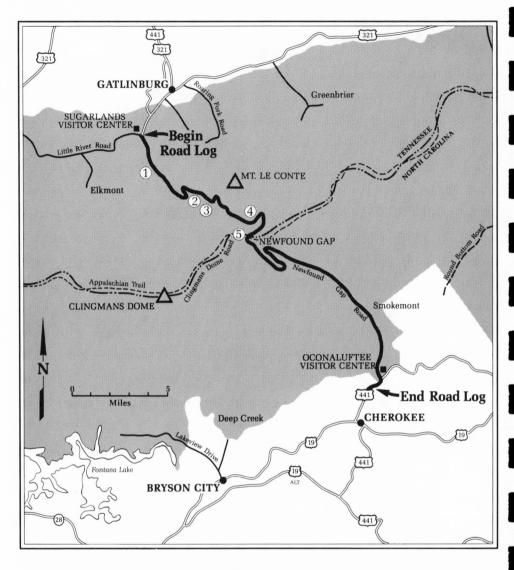

① Mt. Le Conte scenic view, Log Mile 2.1

② Chimney Tops scenic view, Log Mile 5.4

③ Beginning of Hiking Trail to Chimney Tops

④ Anakeesta Ridge landslide area, Log Mile 10.8

⑤ Newfound Gap, Log Mile 12.9

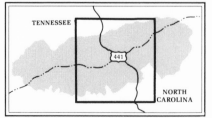

Road Log 2

Route: U.S. 441 from the intersection with Little River Road at Sugarlands Visitor Center in Tennessee to the Cherokee Indian Reservation in North Carolina
Beginning altitude: 1,475 feet
Ending altitude: 1,970 feet
Maximum elevation attained: 5,040 feet
Trip length: 29.9 miles
Nature of the road: Two-lane, paved highway, very curvy, with two short tunnels and some steep sections; occasionally closed due to snow and ice during winter months
Special features: Scenic vistas from within the heart of the Park; crosses Appalachian Trail at Newfound Gap; hiking trail to Chimney Tops; ends at Cherokee Indian Reservation
Hiking trail (optional): The Chimney Tops
For reference: Thunderhead and Anakeesta formations, block fields, faults, basement complex, quartz, landslides (wedge failures)

This route takes you through the heart of the Smokies, providing breathtaking vistas and a glimpse of some of the oldest rocks in the Park. As you travel along U.S. 441 from the Sugarlands Visitor Center toward Newfound Gap, rock units from the Ocoee Supergroup (Class II type rocks) can be observed in bluffs and roadcuts. These sedimentary rocks which have been changed by heat and pressure (metamorphosed) form such topographic prominences as the Chimney Tops, Mt. Le Conte, and Clingmans Dome.

Newfound Gap, with its nearly mile-high elevation, provides stunning vistas into both Tennessee and North Carolina. As U.S. 441 crosses the Appalachian Trail at Newfound Gap, the route quickly descends from the crest of the Smokies into the Oconaluftee River valley. Near the Oconaluftee Ranger Station, a roadcut along U.S. 441 provides an excellent view of the oldest rocks in the Park: these Precambrian rocks are called basement complex (Class I) because of their origin deep within the earth.

Two major faults will be crossed during this road trip. However, neither fault line will be visible. Subtle effects— such as a change in the rock type or the sudden appearance of gray boulders on the slopes along the road—are clues to the presence of these faults.

The Greenbrier fault is crossed twice along this route, first at Log Mile 1.7 at the base of the Smokies in Tennessee and then at Log Mile 27.9 near the Oconaluftee Ranger Station in North Carolina. The Greenbrier fault, a major structural element of the Smokies, is found at the surface in the eastern half of the Park; along this fault the basement complex rock has been brought into contact with the metamorphosed sedimentary strata of the Ocoee Supergroup.

A second fault, the Oconaluftee fault, is crossed at Log Miles 13.6, 17.9, and 21.6 as U.S. 441 winds its way down the Oconaluftee River valley, which is underlain by the fault. This valley is easy to see as you drive into North Carolina just past Newfound Gap.

Noteworthy aspects of this road trip include the dominant exposures of the Thunderhead and Anakeesta Formations, which form the crest of most of the peaks. Many popular hiking destinations—including Mt. Le Conte, the Chimney Tops, Charlies Bunion, the Appalachian Trail, and Clingmans Dome (all of which are accessible from this road trip)—have their origin directly related to the underlying geology. Numerous rock exposures in both roadway cutslopes and natural outcrops are visible along the route. Tilted bedrock,

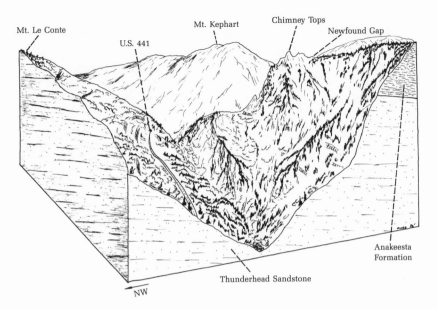

Figure 10. *The massive Thunderhead Sandstone and the slate/meta-siltstone Anakeesta Formation provide the subsurface character in the Chimney Tops section of the Park. The junction between the Anakeesta Formation and the Thunderhead Sandstone is located approximately a quarter of the way down from the top of the Chimneys.*

rusty iron staining of the pyrite-containing Anakeesta Formation, and exposures of the granitic gneiss basement complex can all be observed along this route.

Of geomorphic importance are the large deposits of broken rock and debris (talus and colluvium) that cover many of the mountain slopes in the Chimney Tops area. This unstable material—piles of gray boulders that appear to flow down the mountainside—was produced by the weathering of the massively bedded Thunderhead Formation during the Pleistocene age.

From Log Mile 10.8 to 11.5, the route crosses an area prone to landslides. These landslides are wedge failures of the bedrock and result in rock and debris flows which sometimes cross the highway. Surface scars of these landslides appear as dark-gray troughs, free of vegetation and exposing flat planes of bedrock. These scars can be seen especially well from an overlook at Log Mile 12.1.

Several dramatic vistas—including Morton's Overlook, Newfound Gap, and Webb Overlook—are found along this route.

Road Log 2

Mileage	Description
0.0	Begin at the intersection of U.S. 441 and Little River Road at Sugarlands Visitor Center; travel south on U.S. 441 toward Cherokee, N.C. The trip starts in an area underlain by tilted strata of Roaring Fork Sandstone of the Ocoee Supergroup.
0.1	On your right, the dark gray rocks forming a small ledge in the slope are exposures of Roaring Fork Sandstone. Notice that the strata are tilted, dipping away from horizontal.
1.0	Quiet walkway on the left.
1.7	At this point you cross the Greenbrier thrust fault, where Elkmont Sandstone and the older Roaring Fork Sandstone are in contact. Although the fault itself is not visible, you can expect to see larger beds of gray rock exposed along the roadway.
2.0	On your right in the curve are exposures of massive beds of Elkmont Sandstone strata.
2.1	Stop at the scenic pulloff on your left for a beautiful view of Bull Head and Balsam Point on Mt. Le

Conte. The view to your left is the Sugarlands Valley. The abrupt change in topography marks the boundary between the valley, part of the foothills section underlain by Roaring Fork and Elkmont Sandstones, and the peaks, underlain by the Thunderhead Formation.

2.2   Quiet walkway on the left.

2.5   Here is the approximate boundary between the Elkmont Sandstone and the overlying Thunderhead Formation. The boundary itself cannot be seen, but you will notice the gray boulders of sandstone talus (or colluvium) which originates from the Thunderhead Formation at a bluff about half a mile above the road. The bluff is visible from here only during the late fall and winter, when the leaves have fallen from the trees.

3.3   Quiet walkway.

3.5   Adjacent to the roadway are exposures of Thunderhead Sandstone, generally gray in color. In these massive exposures the bedding is visible as slightly dipping lines.

3.6   Beginning here is an area of gray, blocky boulders (colluvium) from the Thunderhead Formation.

4.5   Entrance to the Chimneys picnic area.

5.1   The Horseshoe.

5.4   At the scenic pulloff on your right you will get an excellent view of the Chimney Tops (straight ahead) and Sugarland Mountain (large ridge to the right of the valley). The exposures of medium to dark gray rock across from the pulloff are of the Thunderhead Formation; notice the thick, massive bedding and its slight dip to the right (the south-southeast) as viewed from the pulloff.

6.1   A large area of colluvial boulders is on your left; massive cliffs of Thunderhead Sandstone are approximately one-fourth mile above the road and easily visible during the fall and winter.

Along U.S. 441 on the Tennessee side of the Smokies, numerous large boulders and blocks of colluvium appear, beginning near Log Mile 3.6. These block fields and block slopes are actually talus material derived from weathering of the Thunderhead Sandstone.

At Log Mile 5.4 a scenic view of the Chimney Tops also provides a fine opportunity to observe the result of weathering of the Anakeesta Formation.

Between Log Miles 5.4 and 6.5, excellent exposures of massive bedded Thunderhead Sandstone can be seen. Note the bedding (layering) of the rock strata and also its attitude (dip), indicating movement from its original horizontal position.

6.8 Tunnel.

7.0 Parking area and trailhead for the Chimney Tops trail (four miles roundtrip) are on the right (see Hiking Trail 2, pp. 103–9).

7.5 Loop in the road.

8.1 As you cross over the boundary into the Anakeesta Formation, notice the change in topography, from steep to flatter slopes.

8.8 Parking area on the left for Alum Cave Bluffs and Mt. Le Conte hiking trail.

9.4– During this part of the route you will cross a small
9.7 slice of exposed Thunderhead Sandstone strata; however, it is readily visible only in the small roadcut on the right side of the main road.

9.7   Here you reenter an area underlain by the Anakeesta Formation (continuing from this point to Log Mile 13.6).

10.8   On your left above the road is an area prone to wedge-failure landslides.

11.2   On your left is the site of an extensive landslide; notice the wedge-shaped topographic scar going up the drainage ravine to the left of the road.

11.5   Now you begin traversing the steep incline to Newfound Gap. The rock exposures you see along the road belong to the Anakeesta Formation: these thin, flat rocks have a slaty character and numerous fractures; their rusty color comes from oxidation of the mineral pyrite which they contain.

12.1   The scenic pulloff on the right provides an excellent opportunity to view a landslide area along the Anakeesta Ridge. The long, narrow gray patches on the mountain slope are landslide scars.

At Log Mile 12.1 a pulloff on the right side of the road provides an excellent view of several landslide scars on Anakeesta Ridge.

The abundance of spruce and fir trees is typical of the change in vegetation that takes place above an elevation of 4,500 feet.

12.3   Tunnel.

12.6   Morton Overlook. Use your binoculars for enjoying the scenic view down the valley of the West Prong of the Little Pigeon River; if you look closely, you can see the Chimney Tops along the left side of the valley.

12.9   Newfound Gap parking area. The large cutslope along the parking area exposes rusty, grayish-brown rocks with a flat surface; these are slate and metasiltstone of the Anakeesta Formation. The metasiltstone appears blocky, and the slate looks like thin, flat slabs. Notice that the cutslope lacks vegetation. The reason is that a naturally occurring sulfuric acid solution, produced by oxidation and breakdown of pyrite undergoing weathering, sterilizes the slope.

13.1   Clingmans Dome Road is on your right. The large roadway cutslope ahead on the right of U.S. 441 was the site of a landslide during the summer of 1985. The landslide resulted when rock supporting the

Morton Overlook (Log Mile 12.6) provides a fine view down the valley of the West Prong of the Little Pigeon River. Along the ridge crest, in the upper left of the photograph, are the Chimney Tops.

This view from Morton Overlook shows the valley of the West Prong of the Little River cloaked in winter snow. (That is U.S. 441 winding along the valley floor.)

From Newfound Gap, nearly a mile high, is a view down the Oconaluftee River valley into North Carolina and in the distance more of
the Blue Ridge province. (U.S. 441 is visible along the right.)

Numerous exposures of the Anakeesta Formation can be observed
along U.S. 441 in the Newfound Gap vicinity. The characteristic rusty
color of the outcrops results from oxidation of the iron sulfide minerals found in this kind of rock.

slope was removed by highway construction. This undercutting, which exposed the bedding and fracture planes, permitted the rock above to slide down into the road.

Now you begin descending the crest of the Smokies.

13.6 In the middle of the curve you cross the Oconaluftee fault, along which the Oconaluftee Valley (on your left) has formed. The fault, located along the valley floor, is not visible. During road construction in the 1960s, the headwaters of the Oconaluftee drainage were severely damaged by acid runoff from pyrite-bearing rock from the Anakeesta Formation.

13.6– This stretch of road crosses a small section of Thun-
14.0 derhead Sandstone, which appears grayer and more blocky than the other rock.

14.0– Along this section of road the Anakeesta Formation
14.5 is easy to see: again notice the rusty color of the rock.

14.5 Here you leave the Anakeesta Formation and resume Thunderhead Sandstone (brownish-gray, rounded, and massive).

15.2 A scenic view is on your right: the valley below is the Deep Creek drainage area, underlain predominantly by the Thunderhead Formation.

15.9 Webb Overlook on your right provides another vista.

16.1 Thunderhead Sandstone is exposed in a roadcut along the curve; notice the dipping layers of brownish-gray rock.

17.9 Once again you cross the Oconaluftee fault as the road winds down the mountain slopes.

17.9– The road crosses a small area underlain by the Ana-
18.2 keesta Formation (not visible); Richland Mountain is on the left.

21.6 For a third and last time you cross the Oconaluftee fault.

23.6 Collins Creek Campground turnoff.

The Oconaluftee fault, crossed at Log Mile 13.6, is found along the base of the Oconaluftee River valley.

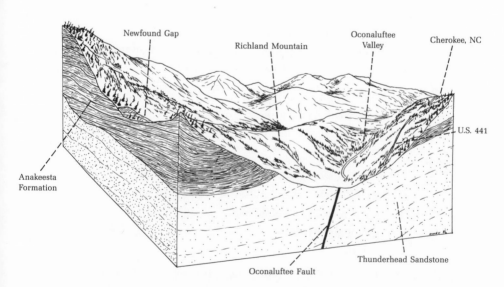

Figure 11. The Oconaluftee fault is the principal influence on the topography of the Oconaluftee Valley.

In a sharp curve on U.S. 441 at Log Mile 16.1 are exposures of Thunderhead Sandstone that illustrate its layered structure and its resistance to weathering.

24.8   The flat area along the road is the old floodplain of the Oconaluftee River.

25.4   Main entrance to Smokemont Campground (on the left).

26.5   Tow String Road (on the left).

27.9   Here you cross the Greenbrier fault. Although the fault itself is not visible, you can see its results: the rocks are more fractured and contain numerous veins of milky-white quartz. This is the beginning of an area underlain by the Precambrian age basement complex (Class I rocks).

28.1   Mingus Mill parking area (to the right).

28.2   Exposures of basement complex bedrock appear on your right in a small roadcut. This dark-gray rock contains many irregular zones of white quartz

An area underlain by the Precambrian age basement complex begins near Log Mile 28. The Oconaluftee River has developed a wide alluvial floodplain which overlies the bedrock.

Exposures of the Precambrian age basement complex are found along U.S. 441 at Log Mile 28.2. The basement complex is composed chiefly of gneiss, schist, and granite; the rocks shown here are granitic gneiss.

masses. Notice the massive character of the rock and its lack of bedding planes: this is a metamorphic rock known as granite gneiss, which resulted from the deformation of a granite rock by heat and pressure.

28.4   Entrance to Job Corps Center (on the left).

28.5   The Oconaluftee Visitor Center on your left is located on the old floodplain of the Oconaluftee River. The large ridge in the distance is Rattlesnake Mountain, which is underlain by the basement complex and a quartz sandstone known as Longarm Quartzite.

29.3   Blue Ridge Parkway (on the left).

29.7   Leave Great Smoky Mountains National Park.

29.9   Enter Cherokee Indian Reservation. The rest of the route of U.S. 441 to the intersection with U.S. 19 is underlain by the basement complex. The area is commercially developed.

End of Road Log 2

Clouds often shroud the Chimney tops. (The rock strata are of the Anakeesta Formation.)

Hiking Trail 2: The Chimney Tops

Location: Trail begins along Newfound Gap Road (U.S. 441)
7.0 miles south of Park Headquarters in Tennessee
U.S.G.S. 7½ minute quadrangle map: Mt. Le Conte,
Tennessee
Start: Chimney Tops parking area, 3,400 feet
End: Chimney Tops, 4,750 feet
Nature of trail: Unpaved foot trail; maximum elevation gain,
1,350 feet
Distance: 2.0 miles

Breathtaking panoramas and exposures of tilted rock strata
make the two-mile hike to the Chimney Tops well worth the
effort. The Chimney Tops gets its name from the nearly vertical holes which have developed through the jutting rock exposure, forming a natural chimney. The trail crosses the West
Prong of the Little Pigeon River and then crosses the Road
Prong three times. Numerous small cascades and rapids
along Road Prong make the first half of the hike very enjoy-
able. The last half of the trail, though somewhat steep, re-
wards you with stunning views from the Chimney Tops. Notable geologic features of this trail include the numerous exposures of Thunderhead Sandstone (medium-gray and banded)

Outcrops of Thunderhead Sandstone are exposed in the creek bed where the Chimney Tops trail crosses over the West Prong of the Little Pigeon River. The rock strata dip upstream.

and the Anakeesta Formation (rusty-colored and shaly). This is a particularly lovely hike during October, when fall colors are at their peak.

Trailhead. The Chimney Tops trail begins along the Newfound Gap Road (U.S. 441) in Tennessee (Road Log 2, Log Mile 7.0) at the Chimney Tops trail parking area. As this is a very popular trail, especially during summer and fall, parking can be a problem. Additional parking spaces, however, can be found a short distance up the road toward Newfound Gap.

Trail details. Upon leaving the parking lot, the trail descends and crosses the West Prong of the Little Pigeon River. As you cross the footbridge, notice the exposures of continuous masses of smooth, light-gray rock beneath the boulders in the stream: these are in-place Thunderhead Sandstone strata. The large, rounded, gray boulders in the stream originated as angular blocks of colluvium that were transported down the mountain from the block fields by mass wasting and stream waters.

A short distance past the first footbridge, the trail crosses over Road Prong near its junction with the West Prong of the Little Pigeon River. As you look upstream from the footbridge, the many layers of banded rock you see exposed in the stream channel are again strata of Thunderhead Sandstone.

Just past the Road Prong crossing, the trail turns to the left and you begin a short, steep grade. At about 0.2 mile you recross Road Prong and begin a moderate climb of approximately 0.5 mile. During this stretch of trail, you cross strata of the Anakeesta Formation, but unfortunately they are not easily seen.

About 0.75 mile from the beginning of the trail, you again cross Road Prong and have a good view of the steeply dipping rock strata of the Thunderhead Formation below the footbridge. About 100 yards past the footbridge, the trail cuts sharply to the right. Here the Indian Gap trail continues to the left, up the Road Prong drainage to the crest of the Smokies.

At the second creek crossing, strata of Thunderhead Sandstone (exposed in Road Prong) form numerous small cascades and falls.

Where the trail approaches the base of the Chimney Tops, strata of the Anakeesta Formation weather to develop a thin, acid soil, requiring the vegetation to compete for root space. Repeated foot traffic has left the root mat bare.

A view of the Chimney Tops from the base of the jutting spire of rock which forms the first chimney. The rock strata dip toward the viewer at about 50°.

To get to the Chimneys, however, you should continue right, past the Indian Gap trail. The path now becomes rocky, and at about 1.25 miles you begin a very steep incline. At this point the trail is very close to the Mingus fault, as you can tell from the nearly vertical dipping strata of the Anakeesta Formation along the left side of the trail. The milky-white boulders are composed of the mineral quartz.

At a switchback in the trail, the grade becomes less steep. Near the crest of the ridge, the vegetation changes to hemlock and red spruce with an undergrowth of rhododendron and mountain laurel. At the top of the ridge, you continue on the main trail over the crest and past a narrow footpath that cuts back to the left. The main trail passes over exposures of Anakeesta strata as it makes a final approach to the base of the Chimney Tops. Now the trail quickly narrows along the crest of a ridge where you can observe tree roots competing for a foothold in the thin soil.

At 2.0 miles the trail reaches the Chimney Tops, where exposures of the Anakeesta Formation jut fifty feet skyward. A

From the crest of the Chimneys is a view of the "backbone" of bedrock formed by the weathering away of strata along joint fractures.
The remaining joint block is exposed as a ridge of rock.

second rock pinnacle is just below and to the north of the first chimney. The trail crosses over the course ("strike") of the dipping strata, where weathering along vertical joints has left a backbone ridge of bedrock exposed; as you face the Chimney Tops from the trail, the dip of the bedrock is toward you. (The flat, smooth planes of rock dipping toward you are the bedding planes.)

From the top of the Chimneys, you can see Sugarland Mountain to the left (west) as well as the valley of the West Prong of the Little Pigeon River and U.S. 441; Balsam Point on Mt. Le Conte is straight across from you (north), and Anakeesta Ridge to your far right (east); behind you (south) is the crest of the Smokies, with Mt. Collins, Indian Gap, and Mt. Mingus.

Caution: Climbing on the rock exposures can be hazardous, especially during inclement weather.

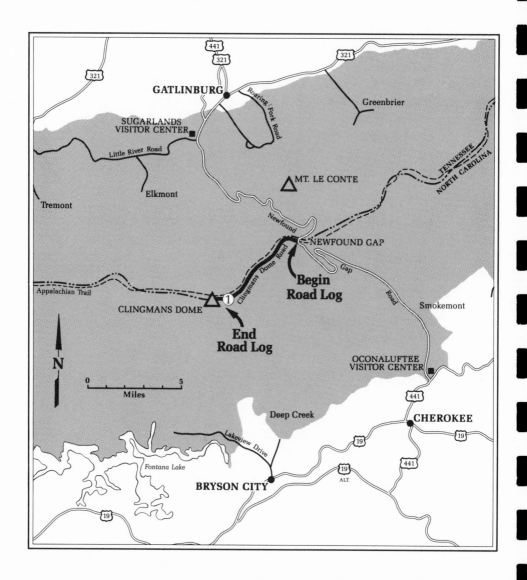

① Beginning of Hiking Trail to Clingmans Dome

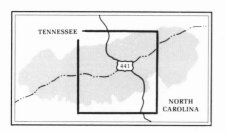

Road Log 3

Route: Clingmans Dome Road from U.S. 441 at Newfound
Gap to Clingmans Dome parking area
Beginning elevation: 5,040 feet
Ending elevation: 6,311 feet
Maximum elevation attained: 6,311 feet
Trip length: 7.2 miles
Nature of the road: Two-lane paved road, very curvy, with
some steep sections; road closed during winter months
Special features: Spectacular views of North Carolina and
Tennessee from the crest of the Smokies; spruce-fir forest;
excellent exposures of Thunderhead Sandstone
Hiking trail (optional): Clingmans Dome observation tower
(elevation 6,643 feet)
For reference: Faults, Thunderhead and Anakeesta forma-
tions, Ocoee Supergroup

Traveling along this route, you will see some of the highest
peaks in the Park, travel its entire route above the mile-high
elevation, and experience the flora, fauna, climate, and geology of the high country. The route begins at Newfound Gap
(nearly a mile high), traversing metamorphosed sedimentary
rocks and climbing through a "Canadian" spruce-fir forest.

The route continues upward, passing near an igneous rock intrusion, providing breathtaking views of the valleys below, and ending at an elevation of nearly 6,200 feet at the parking area for the Clingmans Dome trail.

The rocks in this section of the Park are in the Class II category (see "A View of Geologic History" above), being metamorphosed sedimentary rocks of the Ocoee Supergroup and dating from the Precambrian age. As seen along this route, the rock strata are tilted at varying degrees and exposed at numerous places. Two formations exposed and easy to see include the slaty and pyritic Anakeesta Formation and the massive Thunderhead Sandstone. You should notice the distinct changes in the vegetation and rock exposures as the geology changes from the slaty lithology to the thick-bedded sandstone and conglomerate strata.

Approximately 1.3 miles into this trip, the route crosses the Oconaluftee fault, forming a distinctive topographic feature. Looking east and down into North Carolina from Indian Gap, you can see the obvious long, linear Oconaluftee River valley, which is underlain by the fault.

Important features to look for along this road trip include the massive nature of the Thunderhead Formation and the distinct slaty bedding and rust color of the Anakeesta Forma-

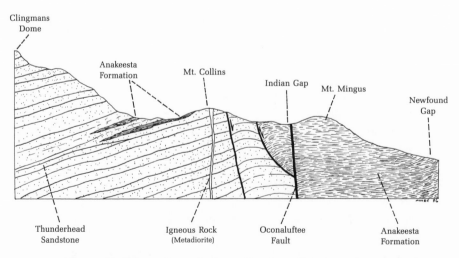

Figure 12. *Schematic geologic profile along Clingmans Dome Road.*

tion. You should also take note of how the overlying soil cover and resulting vegetation respond to the two different rock types.

The noteworthy weathering characteristics of Thunderhead Sandstone can best be observed at the Clingmans Dome parking area. There the strata are exposed in massive, rounded clumps of rock near the beginning of the trail to the top of Clingmans Dome. (At the end of the paved trail, the observation tower atop Clingmans Dome affords a dramatic 360° panorama of the Park.)

Road Log 3

Mileage	Description
0.0	Begin at the intersection of U.S. 441 and the Clingmans Dome Road at Newfound Gap. At this intersection are extensive exposures of shale-like, brown-to-gray rocks of the Anakeesta Formation. The rusty color—iron staining characteristic of the Anakeesta Formation—is due to the oxidation of pyrite, an iron sulfide mineral found throughout this formation. The next 1.8 miles are good for observing the Anakeesta Formation: notice that the rock is shaly, thinly bedded, fractured (with joints and cleavage), and broken up; the soil cover is relatively thin.
1.3	Indian Gap. The trace of the Oconaluftee fault passes through Indian Gap. This gap served as a major crossing and trail for the Indians and early settlers. Take a look at the historical exhibit in the parking area.
1.8	At this point you start seeing massive exposures of thick-bedded, light-gray sandstone strata of the Thunderhead Formation (and leave the Anakeesta Formation). Notice that the outcrops are rounded.

The first 1.8 miles of the Clingmans Dome road (beginning at New-found Gap) is underlain by the Anakeesta Formation, which is exposed in numerous roadcuts. This rock has a shaly character and interbeds of metasiltstone.

2.8   At the pullover on your left is an exhibit about the spruce-fir forest in which you are traveling.

3.7   An access trail to the Appalachian Trail is on the right. A small igneous dike—a thin, tabular intrusion of igneous rock into the surrounding bedrock—cuts through the Thunderhead Formation in this area. Unfortunately, exposures along the road are covered with soil and vegetation, so the dike is impossible to see.

4.7–   The route crosses a small section of the Anakeesta
4.9   Formation. Again notice the rusty color and thin beds of these slaty rock strata in the road cuts.

5.6   The scenic overlook on your left provides an excellent view of the Blue Ridge Province and the eastern half of the Great Smoky Mountains National Park.

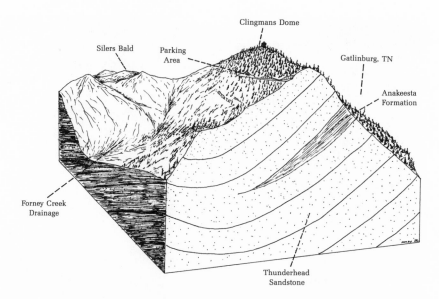

Figure 13. In the Clingmans Dome area of the Smokies, lenses of the slaty Anakeesta Formation interfinger with the massive beds of the Thunderhead Sandstone. (North Carolina is to the left of the drawing and Tennessee is to the right.)

The massive bedded and cliff-forming Thunderhead Formation consists of a coarse-grained metasandstone with thin interbeds of slate and phyllite. Near Log Mile 6.0 extensive roadway cutslopes expose strata of Thunderhead Sandstone.

Along the border of the Clingmans Dome parking area are numerous exposures of the Thunderhead Formation. In this view the bedding dips toward the observer; note the vertical quartz-filled fractures.

The coarse grain of the bedrock is easily observed. The coarser material above is composed of the mineral quartz, which is very stable and resistant to weathering processes.

Above the Clingmans Dome parking area are numerous natural outcrops of Thunderhead Sandstone. The very thin (and in places nonexistent) soil cover attests to the sandstone's ability to resist weathering.

The Clingmans Dome parking area offers a view of a great expanse
of the Blue Ridge province in North Carolina, including a large portion of the Smokies.

7.1   As you enter the Clingmans Dome parking area, observe the exposures of massively bedded sandstone strata of the Thunderhead Formation.

7.2   Along the paved trail to the restrooms you will see numerous outcrops of the Thunderhead Formation. The climb to the observation tower (1.2 miles round-trip) is rewarding (see Hiking Trail 3, pp. 121–25).

End Road Log 3

Clouds cover the observation tower atop Clingmans Dome, where on a clear day a 360° panorama of the Park may be enjoyed.

Hiking Trail 3: Clingmans Dome Observation Tower

Location: At end of Clingmans Dome road (7 miles from
Newfound Gap), along crest of Smokies
U.S.G.S. 7½ minute quadrangle map: Clingmans Dome,
165–SW
Start: Forney Ridge parking area, 6,311 feet
End: Clingmans Dome, 6,643 feet
Nature of trail: Paved foot trail; maximum elevation gain,
332 feet
Distance: 0.6 mile

From the Clingmans Dome observation tower you will get
breathtaking vistas of the Park, including Chilhowee Mountain, Sugarlands Valley, Mt. Le Conte, Newfound Gap, the
Tennessee–North Carolina border, Smokemont, Thomas Divide, and portions of Fontana Lake.

Geologically, Clingmans Dome is particularly interesting
for the metasandstone and slate of the Thunderhead Formation with which it is underlain. Excellent exposures of Thunderhead Sandstone are found in the Forney Ridge parking
area as well as at the beginning of the trail to the observation tower.

In most places on Clingmans Dome, trees struggle to gain a foothold
among the sandstone outcrops.

Trailhead. To reach the beginning of the paved foot trail to the Clingmans Dome observation tower, you take the Clingmans Dome road from Newfound Gap to the Forney Ridge parking area (Road Log 3, Log Mile 7.1); note that this road is closed during the winter months after the first significant snowfall.

The paved trail to the observation tower begins at the west end of the parking area; restrooms are available at the trailhead.

Trail details. At the beginning of the trail, the numerous exposures of smooth, rounded, light-gray rock are Thunderhead Sandstone. In the sandstone exposures adjacent to the trail, you can see several one-half- to one-inch-wide veins of milky quartz. These veins are old fractures that have been "healed" by the infilling of quartz. In the area around the restrooms, numerous rounded rocks are exposed. Here the trees, com-

The once horizontal layers of sandstone along the trail to Clingmans Dome have been tilted to produce steeply dipping strata. (Note the camera case; the rock strata dip to the left.)

Near the upper end of the trail to the observation tower at Clingmans Dome, outcrops of more shaly strata of Thunderhead Sandstone illustrate the cleavage pattern so often developed in rocks of the Smokies.

peting with other plants for a foothold, strive to find organic nutrients within the crevices between each rock outcrop.

Along the right side of the trail, approximately 300 yards from the beginning, you can see more exposures of Thunderhead Sandstone. These outcrops show how the strata dip into the ridge; vertical cracks (joints) can also be observed at this point.

In the outcrops near the spruce-fir exhibit (located about halfway to the tower), you will notice a significant change in the bedding. Below the exhibit the rock dips away from you; above the exhibit the rock dips toward you.

Near the top of the incline, the trail passes some exposures of light-brown, slaty rock, which are a more shaly phase of Thunderhead Sandstone. In this rock you can observe examples of fracture cleavage (cracks forming sharp angles with each other).

On top of the hill, a narrow footpath leads to the Appalachian Trail.

At 0.6 mile the paved trail reaches the base of the observation tower. Displays along the viewing rail at the top of the tower illustrate and label major features of the 360-degree panorama.

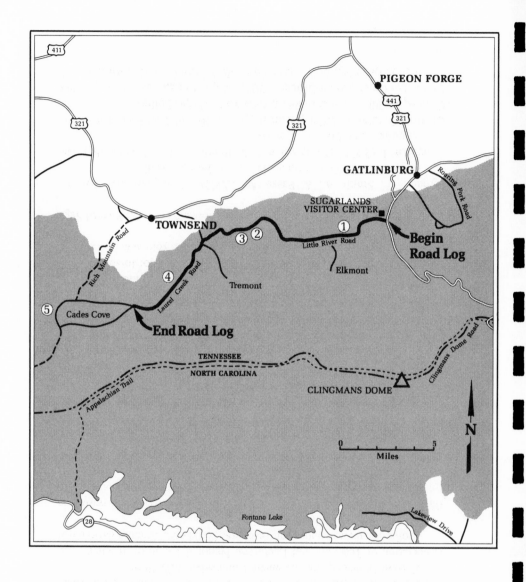

① Maloney Point Overlook, Log Mile 3.5

② The Sinks, Log Mile 11.8

③ Meigs Falls, Log Mile 12.9

④ Trail to Whiteoak Sink, Log Mile 21.3

⑤ Beginning of Hiking Trail to Abrams Falls

Road Log 4

Route: Little River Road from Sugarlands Visitor Center at
Park Headquarters to Cades Cove via the Laurel Creek
Road
Beginning elevation: 1,475 feet
Ending elevation: 1,920 feet
Maximum elevation attained: 2,290 feet
Trip length: 25.3 miles
Nature of the road: Two-lane paved highway, with numerous
curvy sections and moderate grades
Special features: Trailhead to Laurel Falls; "The Sinks";
Cades Cove; numerous exposures of deformed rock strata;
gorge of the Little River
Hiking trail (optional): Abrams Falls
For reference: Geologic windows, faults, cleavage

On this road trip you will travel through a section of the
Great Smoky Mountains National Park containing both metamorphosed Precambrian rocks (altered in their composition
by heat and pressure) and sedimentary Paleozoic limestone.
The Precambrian rocks have been complexly folded and
faulted. In addition, the folded metamorphic rocks along this
route have been thrust-faulted over the younger Paleozoic

limestone strata. Weathering of the rock strata down through the fault plane has in places exposed the younger limestone strata. These geologic "windows" are so called because they permit a look at the younger underlying limestone through the older surrounding rock. Those windows, expressed topographically as coves, are fertile farmlands surrounded by high mountain ridges. Cades Cove is one of several such geologic windows.

On the first part of the road trip you can view most of the Little River gorge and observe its associated vegetation. Massive exposures of bedrock are visible along the gorge, both in the river channel and along the mountain slopes next to the roadway.

The second part of the trip traverses the West Prong of the Little River by way of Laurel Creek Road and crosses Crib Gap, taking you into Cades Cove, one of the most beautiful areas in the Park. Here you get to view geologic windows, where younger limestone strata are exposed.

Traces of the Greenbrier thrust fault, with associated smaller fault blocks, and the Great Smoky thrust fault will be crossed during the trip, though they are obscured from view by soil and vegetation. Exposures of the Great Smoky fault will be easily seen in Cades Cove, however, where younger limestone strata lie in contact with the surrounding older metamorphosed rocks of the Ocoee Supergroup.

Features of this road trip include massive exposures of the dark-gray Thunderhead Sandstone and the gray, shaly Metcalf Phyllite which are found along the Little River gorge. In addition, you will have beautiful scenery: the Little River gorge itself, Cades Cove, and views of the western half of the Smokies crest from Cades Cove.

At the end of this road trip is a description of a hiking trail to Abrams Falls, with pertinent geologic features noted.

Road Log 4

0.0   Leave Sugarlands Visitor Center at the Park Head-
quarters (intersection of U.S. 441 and Little River
Road). Proceed on Little River Road toward Cades
Cove.

For the next twelve miles the road traverses a
complex arrangement of the Precambrian rock strata.
These rocks are metamorphosed sedimentary strata
that have been complexly folded and faulted. The
rock units consist of metasandstone, metasiltstone,
slates, and phyllite. Approaching Fighting Creek Gap
(Log Mile 3.8), you will see the grayish-brown rock
strata of the Pigeon Siltstone.

3.5   Maloney Point Overlook. Here you look out over the
Sugarland foothills, which are primarily underlain by
the Roaring Fork Formation. Although the rock itself
cannot be seen, the effects of its weathering are ap-
parent in the rolling ridges of the foothills. Tops of
the existing foothills would have been part of the
surface of the old valley floor (see "Cenozoic Depos-

Maloney Point Overlook at Log Mile 3.5 provides a good view of the Sugarlands Valley, including the foothills of the Smokies. To the extreme right is Mt. Le Conte.

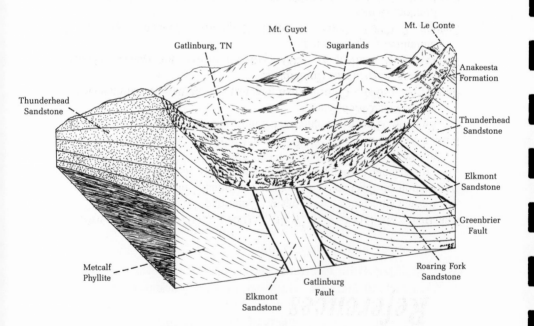

Figure 14. Beneath the Gatlinburg-Sugarlands area the majority of the bedrock dips to the right (southeast), as do most major faults in the Smokies.

its," pp. 36–45). The high peak to the extreme right is Mt. Le Conte.

3.8   Fighting Creek Gap. The popular trail to Laurel Falls begins here. This hike (three miles roundtrip) provides an excellent opportunity for viewing the massive beds of Thunderhead Sandstone over which a small mountain stream flows, forming Laurel Falls.

5.0   Entrance to Elkmont (on the left). The Elkmont area was the site of a major lumbering company before the Great Smoky Mountains National Park was founded; some of the original settlements still remain.

Elkmont Sandstone takes its name from this section of the park and is typically found there, though it also occurs in other locations.

6.3   Here you enter an area which is underlain by Thunderhead Sandstone; it is exposed and easily visible along the channel of the Little River. Many of the massive bluffs throughout the Park are composed of

The hiking trail to Laurel Falls begins at Log Mile 3.8 (Fighting Creek Gap). Metasandstone strata of the Thunderhead Formation are exposed along the trail and form the precipice over which Laurel Falls pours.

Thunderhead Sandstone, and it is typically this formation over which waterfalls, such as Laurel Falls and Rainbow Falls, have developed. If you look closely at the cobbles and boulders in the stream, you will see that their upper edges are inclined in the direction of the current; this effect is called imbrication.

9.1   At this point you cross the Greenbrier fault trace, but it is not generally visible from the road.

9.8   Metcalf Bottoms picnic grounds. This area is underlain by Metcalf Phyllite, much less resistant than Thunderhead Sandstone. Accordingly, the topography in this area is less extreme.

10.4   After recrossing the Greenbrier fault (again not visible from the road), you are back onto the Thunderhead Formation and now enter Little River Gorge. Notice the abrupt change from flat ground to mountain slopes.

11.8   The Sinks. This is a river cascade developed on beds of the Thunderhead Formation which have been turned on end.

A popular swimming area in the Smokies is the Sinks (Log Mile 11.8), formed where the Little River flows over very resistant tilted strata of the Thunderhead Formation.

In this area roadway embankments have had to be reconstructed many times because of the erosive action of the Little River. As these roadway fills have suffered numerous slumps and slides, they have needed frequent repair (note the fill of stacked quarried limestone blocks along the river's edge).

12.9 Meigs Falls Overlook. As you might expect by now, the falls are developed over rock of the Thunderhead Formation.

13.6 Now you cross the Greenbrier fault again and enter an area underlain by the Metcalf Phyllite Formation. Although the fault is not visible, the Metcalf Formation is: a thin-bedded slate and phyllite with a shaly appearance.

17.6 Proceed straight ahead on Laurel Creek Road to Cades Cove; S.R. 73 turns off right to Townsend. This intersection of the roads and the streams is known locally as the "Y".

17.8 Road to Tremont (on the left).

17.85 Here you cross an unnamed fault (related to the Greenbrier fault and not visible from the road). At this point you begin to see exposures of Cades Sandstone, the large beds of brownish-gray rock alongside the road.

17.9 In the roadcut on your left is a landslide scar.

18.0 Here you get a good look at Cades Sandstone, with its deformed beds of rock clearly visible. Turned almost vertical, these beds are gray to brown in color.

18.5 Again you cross the unnamed fault, and again it is not visible. Cades Sandstone ends, and exposures of Metcalf Phyllite begin. Except for a small "window" of limestone at Log Mile 23.2, the road will traverse strata of Metcalf Phyllite from this point to the beginning of the Cades Cove Loop Road.

18.75 Tunnel. Note the strata of light grayish-brown rock around the portal areas of the tunnel. This is the thin-layered Metcalf Phyllite.

Cades Sandstone, which forms the bulk of Rich Mountain and Cades Cove Mountain, is very resistant to weathering and thus has very little soil cover.

At Log Mile 18.75 the Laurel Creek Road tunnels through tilted strata of Metcalf Phyllite. The strata of Metcalf Phyllite, seen above the tunnel portal, are thin, easily deformed (by folding and faulting), and contain some thin interbeds of metasiltstone.

19.0–   Massive exposures of Metcalf Phyllite.
19.2

19.7   The road crosses a creek twice. The exposures of
Metcalf Phyllite in the roadcut on the right are good
examples of this formation: thin laminated bedding
with fractures and small folds. If you wish to stop
and take a close look at this rock formation, there is
a parking area across the road (on your left).

21.2   The trailhead on the left leads to Bote Mountain
Road (see also Log Mile 23.3 below) and Spence
Field at the crest of the Smokies.

21.3   The trailhead on the right leads to Schoolhouse Gap
at the Park boundary. Located near Schoolhouse Gap
is Whiteoak Sink, a large limestone sinkhole (and
geologic window), where several streams empty into
cave openings. (This makes a good hike, approximately
five miles roundtrip. However, you should plan in ad-

*The shaly-appearing Metcalf Formation contains very thin laminae of
phyllite and slate which were deformed during the formation of the
Smokies.*

vance: there are hidden dangers around the cave entrances, and Park permits are required for entering the caves.)

23.2–
23.3   Cross the Great Smoky fault and a window of younger Ordovician age limestone (not visible here). Second-growth timber has obscured the clearing where a settlement was once located to take advantage of the rich limestone-derived soils; this area is known as Sugar Cove.

23.3   The trailhead on the left leads to Bote Mountain Road (see also Log Mile 21.2 above) and onto the crest of the Smokies.

As you begin the steep grade up to Crib Gap, you will notice the change in vegetation from dense hemlock, mountain laurel, and rhododendron to tulip poplar and hardwood forest, free of undergrowth.

24.3   Crib Gap. Now you begin descending into Cades Cove. Notice the shaly exposures of light-brown Metcalf Phyllite on the right side of the road.

25.1   The road on the left leads to the Cades Cove Ranger Station, picnic grounds, camping area, and general store.

25.3   Begin the Cades Cove one-way loop road around the perimeter of Cades Cove. This one-way road, eleven miles in length, traverses back and forth across the Great Smoky fault (generally not visible). The numerous exposures of gray rock along this route are limestone, which weathered to produce the deep, fertile soils tilled by the early settlers in this area. Their log cabins and pastures characterize the cove. Keep your eyes open for the white-tailed deer and wild turkeys commonly seen along this loop road. This is one of the most beautiful sections of the Park and one in which human history is emphasized along with natural history.

You may also be interested in the easy hike (five miles roundtrip) to Abrams Falls which starts approximately halfway around the loop road (see Hiking Trail 4, pp. 139–43).

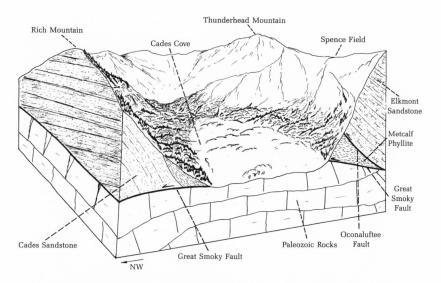

Figure 15. *Cades Cove was formed when erosion penetrated the Great Smoky thrust fault (which lies almost flat) to expose Paleozoic limestones and shales.*

Cades Cove, located in the northwestern part of the Park, is a geologic window, exposing younger limestone strata in the floor of the cove. Cades Cove is surrounded by much older and more resistant strata.

A popular hiking destination is Abrams Falls, located 2.5 miles from the western end of the Cades Cove loop road. The falls are formed by Abrams Creek flowing over very massive beds of Cades Sandstone.

Hiking Trail 4: Abrams Falls

Location: West end of Cades Cove, approximately 5 miles
   along one-way loop road
U.S.G.S. 7½ minute quadrangle map: Cades Cove, Tennessee
Start: Abrams Falls parking area, 1,710 feet
End: Abrams Falls, 1,460 feet (pool elevation)
Nature of trail: Unpaved foot trail; maximum elevation gain,
   280 feet
Distance: 2.5 miles

The Abrams Falls trail follows Abrams Creek as it cuts through
hilly terrain in the northwestern end of the Park. Abrams
Falls and the surrounding area are underlain by a Precambrian rock formation, Cades Sandstone. The trail begins just
inside the Cades Cove geologic window, where the valley
floor is underlain by Paleozoic limestone. Noteworthy features of this trail include the alluvial floodplain and the
gorge of Abrams Creek, the weathering and outcrop characteristics of Cades Sandstone, and Abrams Falls itself.

Trailhead. The Abrams Falls trail starts in the western end of
Cades Cove, approximately 5.6 miles from the beginning of
the one-way loop road (Road Log 4, Log Mile 25.3). After

driving 5.1 miles along the loop road, turn right onto a gravel road with a sign for the Abrams Falls parking area. A 0.5 mile drive ends at the parking area and trailhead.

Trail details. After crossing the footbridge over Abrams Creek, turn left onto the Abrams Falls trail and go along a flat area, which is the narrow floodplain of Abrams Creek. The trail passes through a rhododendron "tunnel" and after 0.5 mile reaches the top of a pine-covered ridge. Here the trail becomes rocky, and numerous exposures of brownish-gray Cades Sandstone can be seen at the top of this ridge. The trail then quickly descends the hill.

After crossing a footbridge over a small tributary stream (notice the dark-gray, blocky outcrops of Cades Sandstone in the stream bottom), the trail again joins Abrams Creek, where you can see the gently dipping bedrock of Cades Sandstone exposed in the creek. After a few hundred yards the trail, becoming more rocky, begins an incline to the crest of Arbutus Ridge and quickly rises above Abrams Creek.

At 1.0 mile the trail crosses Arbutus Ridge, which is underlain by dipping strata of the very weather-resistant Cades Sandstone. If you stand on the trail where it crosses the crest of Arbutus Ridge, you can see parts of Abrams Creek, 200 feet below, as it travels about one mile through an almost complete loop in the gorge.

At the gap on Arbutus Ridge, you can easily see the sandstone bedrock, which dips back to the southeast (towards Cades Cove). The nearly vertical cracks in this light-gray, rough-textured rock are joints, cracks in the bedrock along which groundwater moves and weathering takes place. Because the bedrock weathers to a very thin or, in places, nonexistent soil, it is difficult for plants to establish a foothold.

The trail now quickly descends Arbutus Ridge and at 1.5 miles crosses a footbridge over a small tributary stream. The path of Abrams Creek is paralleled by the trail, which passes through a 0.5 mile stretch of rhododendron tunnel. The thick masses of rhododendron thrive in the moist, acid humus soil along creek banks in the Smokies.

At 2.0 miles the trail again begins to rise above Abrams Creek and again passes over a small ridge. Notice that the vegetation changes from rhododendron and hemlock to moun-

At approximately one mile along the Abrams Falls trail (the horse-shoe turn), exposed strata of Cades Sandstone strike along the ridge and dip to the south.

In-place strata of Cades Sandstone are exposed along Abrams Creek, where erosion has removed all soil cover. Note that the dip of the rock strata is upstream.

Cades Sandstone, seen here at Abrams Falls, has thin to thick beds of sandstone and shale. The more shaley strata have developed cleavage (the characteristic right-to-left diagonal fracture pattern) as a result of stress induced in the rock during deformation of the strata.

tain laurel and pine: this reflects a change from moist humus soil to thin, dry, acid soil. Here the Cades Sandstone is composed of gray sandstone and interbeds of a more shaly rock; the abundant yellowish-brown to reddish soil is derived from the weathering of the more shaly units of Cades Sandstone.

Now the trail quickly descends the ridge and crosses Wilson Branch. At about 2.5 miles the trail to Abrams Falls cuts sharply to the left, away from the main trail. Passing over the mouth of Wilson Branch, the trail enters the Abrams Falls viewing area, where you can easily see the 20-foot-high cascade. The falls are formed over weather-resistant and massive beds of Cades Sandstone.

In the rock wall to the left of the falls, you can get a good look at exposures of the bedrock, which dip upstream at angles of 10 to 20 degrees. The thin layers, near the lower third of the rock wall, are composed of a more shaly and less competent material which fractures more easily. (Competence, in geological vocabulary, refers to the ability of beds or strata, because of massiveness or inherent strength, to lift

Abrams Falls is formed where Abrams Creek flows over a bluff of Cades Sandstone. The strata dip upstream (away from the viewer).

not only their own weight but also that of overlying rock.) This rock develops a feature of cleavage commonly seen in the Smokies area: the diagonal lines within the rocks beds are the planes of cleavage.

Notice that Abrams Creek becomes very wide at the falls: the plunge pool is over 100 feet wide.

Caution: It is not advisable to climb on the rocks around the falls. Because the rocks are covered with algae, they are slippery, particularly when wet.

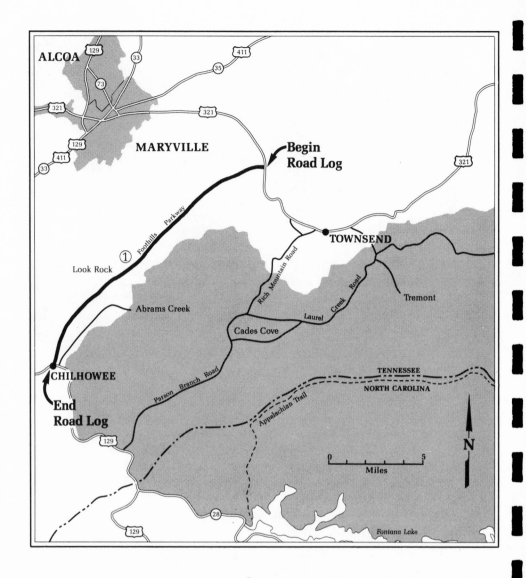

① Beginning of Hiking Trail to Look Rock Tower

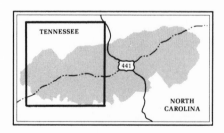

Road Log 5

Route: Foothills Parkway from U.S. 321 at Walland, Tennessee, to U.S. 129 at Chilhowee Lake
Beginning elevation: 1,010 feet
Ending elevation: 890 feet
Maximum elevation attained: 2,570 feet
Trip length: 17.1 miles
Nature of road: Two-lane paved highway along crest of Chilhowee Mountain, with moderate grades and a few curvy sections, but mostly straight to slightly curvy; occasionally closed due to ice and snow in winter months.
Special features: Very scenic drive; numerous overlooks of adjacent Ridge and Valley Province and of western half of Great Smoky Mountains National Park; numerous exposures of Cambrian age sedimentary rocks
Hiking trail (optional): Look Rock observation tower
For reference: Cambrian age rocks, faults, Valley and Ridge province, quartz and calcite minerals

This road trip will give you some of the most maginificent views of the Great Smoky Mountains National Park. Beginning in Miller Cove, which is underlain by limestone, the route goes up and along the crest of Chilhowee Mountain, which is predominantly sandstone and shale. The route descends Chilhowee Mountain, crossing Miller Cove fault, to end at Chilhowee Lake.

You will see both sedimentary and metamorphic rocks along this route. Limestone, sandstone, and shale strata of Cambrian-age formations will be visible for three-fourths of the trip. Along the last part of the route, metamorphosed rocks of the Precambrian Wilhite Formation are exposed, and deformed strata of slate, phyllite, and metasiltstone can be seen adjacent to the roadway.

Features of this route include massive sandstone beds which dip into the roadway; excellent views of the foothills and main crest of the Smokies; views of the adjacent Ridge and Valley province, where younger, less weather-resistant Paleozoic rocks are found; and finally, the deformed metamorphic rocks of the Wilhite Formation.

Road Log 5

Mileage	Description
0.0	Leave U.S. 321 and enter ramp to the Foothills Parkway.
0.15	Enter the Foothills Parkway; turn west, going toward U.S. 129. For approximately the next half mile you will notice the deep orange-red clay soils; these are a weathered residue of the Shady Formation (composed of dolostone).
0.7	Here you cross the boundary between the Shady Formation and the Chilhowee Group Helenmode Formation (a shaly rock), but neither is visible along the road.
1.1	As you cross the boundary between the Helenmode Formation and the Hesse Sandstone, however, you can observe exposures of the latter: in the roadcuts on your right, the light brown to buff-white rock is Hesse Sandstone. Notice the dip of rock into the roadway; this can precipitate block-glide rock falls and landslides, in which the layers of rock simply slide off or glide into the road.

Up Chilhowee Mountain on the Foothills Parkway are dipping strata of the Cambrian age Hesse Formation. Here road excavation has undercut the bedding planes of the bedrock.

1.4   At the overlook on the left you get your first view of the foothills of the Smokies. These low, rolling hills extend between Chilhowee Mountain and the main crest of the Smokies.

3.2   The scenic overlook to the right provides your first view of the adjoining Ridge and Valley province; notice the long parallel ridges and valleys. (Maryville, Tennessee is in the distance.) Unfortunately, on hot summer days thick haze may obscure this view.

3.8   Notice the cutslope on your left: this well-jointed sandstone of the Hesse Formation has a potential for rock falls.

4.6   Along the right side of the road are outcrops of Hesse Sandstone.

5.4   At this point there are scenic overlooks on both sides of the road.

A scenic view from the crest of Chilhowee Mountain (Log Mile 3.2) affords a glimpse of the adjacent and less mountainous Ridge and Valley province.

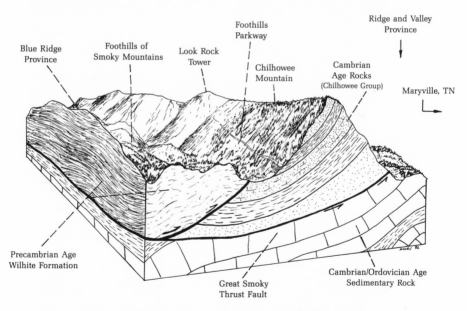

Figure 16. The Great Smoky thrust fault has moved older Precambrian and Cambrian age rocks over younger Ordovician age rocks in the area between Walland and Look Rock. The Great Smoky thrust fault generally marks the boundary between the rolling Ridge and Valley province and the more mountainous Blue Ridge province.

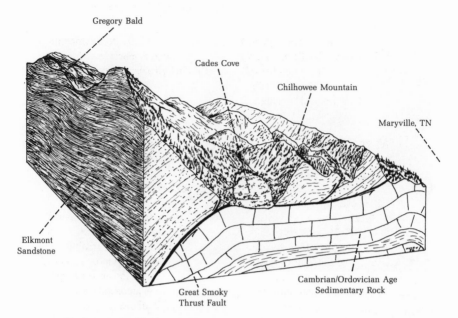

Gregory Bald

Cades Cove

Chilhowee Mountain

Maryville, TN

Elkmont
Sandstone

Great Smoky
Thrust Fault

Cambrian/Ordovician Age
Sedimentary Rock

Figure 17. *In the western portion of the Park, Gregory Bald is located along the crest of the Smokies, while Chilhowee Mountain lies along the border of the Ridge and Valley province.*

5.9   The massive light-brown sandstone and brownish-gray shale outcrops on your right consist of Nebo Sandstone and Murray Shale (part of the Chilhowee Group).

7.0   For the next 1.8 miles, the Parkway travels along the crest of Chilhowee Mountain.

8.8   The Parkway here leaves the crest and moves to the south side of the mountain.

9.3   Cross bridge.

9.45   A road enters on the left from Look Rock Campground and Top of the World Resort.

9.8   Look Rock. Be sure to stop at the scenic overlook on your left. You may also wish to take a side trip to the observation tower; it is only one mile roundtrip and on a paved trail (see Hiking Trail 5, pp. 157–59). The light-brown layered rocks across the road from the parking area belong to the Hesse Formation.

Near Look Rock along the Foothills Parkway, a roadway cutslope has exposed shale strata of the Murray Formation.

The scenic overlook at Look Rock (Log Mile 9.8) offers vistas across the foothills to the crest of the Smokies. A short trail leads from the parking area across the road to an observation tower at the top of the mountain.

The crest of Chilhowee Mountain is upheld by the very weather resistant Hesse Quartzite.

10.0   In the roadcut on your right, Murray Shale is again exposed: the flat planes of brown shale dip toward the road.

10.2   Cross the bridge over a country road. Here is Murray Gap, at an elevation of 2,320 feet.

10.9   Begin the steep descent of Chilhowee Mountain toward U.S. 129.

11.2   On the right you will see Hesse Sandstone outcrops dipping into the road.

13.9   Notice the sandstone exposed in the roadcuts to the right: the bedding lines of the rock vary from vertical to near 30 degrees because of the movement of bedrock by the Miller Cove thrust fault.

14.6   After crossing the Miller Cove thrust fault (not visible), you begin seeing exposures of the Wilhite Formation. The shaly dark-gray rock is composed of

Severely deformed strata of the Wilhite Formation are exposed in road cuts near Log Mile 16.0. Numerous fractures in the bedrock are filled with the minerals quartz and calcite.

phyllite and slate; the thick-bedded brownish-gray to gray rock is sandstone and metasiltstone.

14.8   In the cutslope on your right are small brown chips of phyllite.

15.7   Notice the massive sandstone and metasiltstone exposures of the Wilhite Formation along the road.

16.0   The dark-gray shaly material of the Wilhite Formation is severely deformed (due to movement along the Miller Cove fault), resulting in numerous fractures that have been filled with minerals; these are the white streaks in the rock. This mineral infilling consists of quartz and calcite; the quartz can easily scratch the calcite, but the calcite cannot scratch the quartz.

The Miller Cove thrust fault has moved strata of the Precambrian age Wilhite Formation onto the younger Cambrian age Chilhowee Group. This fault movement has severely deformed the less competent strata of the Wilhite Formation.

At the junction of the Foothills Parkway and U.S. 129 is Chilhowee Lake (Log Mile 17.1), which was formed by damming the Little Tennessee River.

17.1   Junction with U.S. 129 at Chilhowee Lake. This lake was formed by Chilhowee Dam, constructed by Alcoa (Aluminum Company of America) for generating hydroelectric power.

End Road Log 5

Views from the Look Rock observation tower are spectacular: the Ridge and Valley province to one side and the Blue Ridge to the other. Pictured here is the crest of the Smokies, with Mt. Le Conte on the extreme left and Thunderhead Mountain on the right.

Hiking Trail 5: Look Rock Observation Tower

Location: On top of Chilhowee Mountain, along Foothills
  Parkway, in Blount County, Tennessee
U.S.G.S. 7½ minute quadrangle map: Blockhouse, Tennes-
  see, 148 N.W.
Start: Look Rock parking area, 2,460 feet
End: Look Rock observation tower, 2,640 feet
Nature of trail: Paved foot trail; maximum elevation gain,
  180 feet
Distance: 0.5 mile

The 0.5 mile walk to the Look Rock observation tower provides both a look at exposures of Cambrian age sandstone
and fine views of the foothills and the western part of the
Great Smoky Mountains National Park. From the observation
tower, located on the crest of Chilhowee Mountain at an elevation of 2,640 feet, the views are spectacular: to the north of
Chilhowee Mountain, you can see the Ridge and Valley province, and to the south the Blue Ridge province.

Trailhead. The beginning of the trail to the Look Rock observation tower is located along the Foothills Parkway at a parking area overlook (Road Log 5, Log Mile 9.8) approximately

Thousands of years of weathering has reduced these exposures of Hesse Sandstone to rounded masses of rock.

0.3 mile past the turnoff to Top of the World resort. Note: the parkway is occasionally closed because of snow and ice during the winter months.

Trail details. From the upper end of the parking area, follow the signs across the parkway and into the woods. For the first 100 yards of the trail, you pass near gray, rounded clumps of outcroppings of Hesse Sandstone (a sedimentary rock of the Cambrian age). Pine and oak trees dominate on the thin, dry, sandy soils of the Chilhowee Mountain crest. The trail continues winding through the woods, nearly to the crest of the mountain. Here the trail ends at a service road. Turn left and follow the service road for 200 yards to the base of the observation tower. Around the base you can see numerous weatherbeaten outcrops of Hesse Sandstone.

An exhibit at the end of the trail explains some details of the geologic history of the area.

Along the crest of Chilhowee Mountain, the trail to the Look Rock observation tower passes outcrops of Hesse Sandstone which dip southeast toward the Smokies.

From the observation tower you have vistas of Rich Mountain, Clingmans Dome, Thunderhead Mountain, and Gregory Ridge—all of these in the Smoky Mountains. On very clear days you can also see the cities of Maryville and Knoxville to the north of Chilhowee Mountain.

Glossary

Alluvium. Sediments formed by rivers and streams

Amphibolite. A metamorphic rock composed largely of amphibole and feldspar

Anticline. A fold or arch of stratified rocks in which the strata dip in opposite directions from a common ridge or axis

Anticlinorium. A large anticline composed of smaller anticlines and synclines

Aquifer. A porous rock layer from which water may be obtained

Argillaceous. Clayey; containing clay minerals or their metamorphic products

Argillite. Dark, fine-grained rock without cleavage or schistosity, resulting from low-grade metamorphism of claystone or mudstone

Arkose. A sandstone containing at least 25 percent feldspar usually derived from erosion of granitic rocks

Augen structure. "Eyes" or knots of mineral or rock fragments around which foliation of flaser structure is strongly bent

Avalanche. A large mass of snow or ice, sometimes accompanied by other material, moving rapidly down a mountain slope; sometimes applied to rapidly moving landslide debris

Basement. An older rock mass, usually igneous or metamorphic, on which younger rocks have been deposited

Bedding. Layering in sedimentary rocks

Bedrock. Solid rock underlying weathered or transported material

Block diagram. A three-dimensional perspective representation of geologic or topographic features, showing a surface area and generally two vertical cross sections

Block field. A term used to describe the accumulation of rock talus that is wider than it is long; usually found on mountain slopes below sources such as resistant rock bluffs or ledges

Block slope. A contiguous accumulation of rock fragments that is elongated parallel to contour orientations and frequently grades upslope into talus

Block stream. An accumulation of rock fragments (usually boulder size and larger) that is elongated in the slope direction; lengths can range from 30 feet to over 3,000 feet

Boreal. Northern

Boulder. A large, rounded block of stone lying on the surface of the ground or sometimes embedded in loose soil

Calcite. A mineral, calcium carbonate ($CaCO_3$), the principal constituent of limestone

Cambrian. The first period of the Paleozoic Era, from 500 to 600 million years ago

Carbonate. Rocks containing carbon and oxygen in combination with sodium, calcium, or other elements, particularly in limestone or dolomite

Cave. A natural cavity, recess, chamber, or series of chambers beneath the surface of the earth, generally produced by a solution of limestone

Clastic. A term applied to rocks composed of fragmental material derived from preexisting rocks

Cleavage. The tendency for rocks to split along definite planes which generally have no relation to bedding

Colluvium. A general term applied to loose and incoherent deposits, usually at the foot of a slope or cliff and brought there chiefly by gravity (talus and cliff debris are examples of colluvium)

Competent. A term applied to rocks capable of sustaining stress without being greatly deformed

Conformable. Describes strata deposited without significant disturbance or removal of previously deposited strata

Conglomerate. Rock composed of rounded, waterworn fragments of older rock, usually in combination with sand

Cross section. A profile portraying an interpretation of a vertical section of the earth explored by geological and/or geophysical methods

Crystalline. A term applied to rocks composed wholly of crystalline mineral grains; that is, igneous and metamorphic rocks as distinct from sedimentary rocks

Current bedded. Shows bedding features (crossbedding or ripple mark) indicating deposition by currents of water

Debris. Rock and mineral fragments produced by weathering of rocks; synonymous with detritus

Debris flow. A general designation for all types of rapid flowage involving debris of various kinds and conditions

Debris slide. The rapid downward movement of predominantly unconsolidated and incoherent earth and debris in which the mass does not show backward rotation but slides or rolls forward, forming an irregular deposit

Deformation. Change of shape or attitude of a rock body by folding, shearing, fracturing, compression, etc.

Detrital. Composed of detritus or debris

Detritus. See Debris.

Devonian. The fourth period of the Paleozoic Era, from 350 to 400 million years ago

Dip. The angle at which a stratum or any planar feature is inclined from the horizontal

Dolomite. A rock composed essentially of the mineral dolomite or $(CaMg)CO_3$

Downwasting. The diminishing in thickness of glacial ice during natural processes by which a glacier loses substance

Erosion. The process of disintegration and removal of the rocks at the earth's surface by weathering and moving water, wind, ice, or landslide

Escarpment. A cliff or steep slope edging a region of higher land

Fault. A fracture in the earth's crust along which rock on one side has been displaced relative to rock on the other

Fault scarp. A cliff formed by a fault, usually modified by erosion

Fauna. The animal life existing at a particular time or locality

Feldspar. A group of abundant light-colored, rock-forming minerals belonging to the silicate class

Flaser structure. Lenses of granular material separated by wavy ribbons and streaks of finely crystalline foliated material

Flora. The plant life existing at a particular time or locality

Fluvial. Of or pertaining to rivers; produced by river action

Foliation. Parallel alignment of platy mineral grains or flattened aggregates in a metamorphosed or sheared rock

Formation. A distinctive body of some one kind or related kinds of rocks, selected from a succession of strata for convenience in mapping, description, and reference

Geomorphology. A branch of geology which deals with the earth's surface features and landforms

Geosyncline. An elongate depositional basin of continental proportions which is filled by sedimentary rocks over a long period of geologic time

Gneiss. A visibly crystalline metamorphic rock possessing mineral layering or foliation but not easily split along foliation surfaces

Grade (of metamorphic rocks). Refers to the pressure-temperature conditions (low, medium, or high) at which metamorphism occurred

Granite. In the strict sense, a visibly grained igneous rock composed essentially of alkali feldspars and quartz. Commonly, any rock of this composition and texture, whether igneous or metamorphic in origin

Granitic. Pertaining to granite, or similar to granite in composition or texture

Granitization. Metamorphic transformation of nongranitic rocks to granite-like rocks

Granodiorite. A granitic rock in which soda-lime feldspar is at least twice as abundant as potassium feldspar

Groundwater. Subsurface water filling rock pore spaces, cracks, or solution channels

Group. A stratigraphic unit consisting of several formations, usually originally a single formation subdivided by subsequent research

Homocline. Rock strata which are tilted uniformly in the same direction

Hydrostatic pressure. Pressure caused by weight of water in water-bearing rock layers

Igneous. A term applied to rocks formed by crystallization or solidification from natural silicate melts, generally at temperatures between 600° C and 1000° C

Intercalated. A body of material interbedded or interlaminated with another

Interglacial. The time between major advances of continental glaciers

Intrusive. A term for rocks, especially igneous rocks, that have penetrated other rocks

Joint. A fracture in rock along which no appreciable movement has occurred

Kaolinite. A white to tan silicate clay mineral usually occurring in colloidal masses; a secondary mineral resulting from the weathering of feldspar, kaolinite can be found in the park as clay deposits in the soil

Karst. A distinctive type of landscape where solution in limestone layers has caused abundant caves, sinkholes, and solution valleys, often with red soil residue

Klippe. Part of a thrust sheet isolated by erosion from the remainder of the sheet

Laminae. Thin rock layers, generally less than 1 cm thick, of sedimentary or other origin

Limestone. A sedimentary rock composed of calcite, the mineral form of calcium carbonate; the term is also used somewhat loosely to cover both true limestone and dolomite, a very similar rock that includes considerable magnesium in its chemical composition

Magma. A hot, mobile silicate mixture of crystals and melt within the earth's crust

Massive. A term applied to thick bodies of homogenous rock uninterrupted by bedding surfaces, fractures, or other mechanical discontinuities

Mass wasting. A general term for a variety of processes by which large masses of earth material are moved by gravity either slowly or quickly from one place to another

Matrix. The small particles of a sediment or a sedimentary rock which occupy the spaces between the larger particles that form the framework

Megascopic. Visible with the unaided eye or with a hand lens

Metamorphic. Pertaining to or resulting from metamorphism

Metamorphism. The process whereby sedimentary or igneous rocks have been altered by heat and pressure accompanying deep burial in the earth's crust

Mississippian. The fifth period of the Paleozoic Era, from 310 to 350 million years ago

Mudflow. A flowage of heterogeneous debris lubricated with a large amount of water, usually following a former stream course

Nivation. Frost action and mass wasting beneath a snowbank

Orodovician. The second period of Paleozoic Era, from 430 to 500 million years ago

Orogeny. Mountain-building

Outcrop. Any exposure of bedrock

Overthrust. A type of fault in which an extensive slab of rock is moved across a nearly horizontal surface

Paleozoic. The second era of Phanerozoic geologic time, from 225 to 600 million years ago

Pegmatite. Igneous rock of unusually coarse or varied texture occurring in intrusive bodies generally a few feet to a few hundred feet long

Permafrost. Permanently frozen ground (subsoil)

Phanerozoic. Comprises Paleozoic, Mesozoic, and Cenozoic; eon of evident life

Phyllite. A metamorphic rock similar to schist but finer grained, so that the constituent grains cannot be seen with the unaided eye

Plutonic. A term applied to rocks and processes occurring deep within the earth's crust

Precambrian. Geologic time before the Paleozoic Era

Pyrite. A metallic, brass-colored iron ore mineral (FeS_2), often called fool's gold, used as a source of sulfur for sulfuric acid

Quartz. A hard, glassy mineral, silicon dioxide (SiO_2) that is one of the commonest rock-forming minerals, and a silicate group mineral

Quartzite. A sedimentary or metamorphic rock composed largely of quartz grains cemented by silica

Radioactive. A term applied to minerals or rocks containing atoms whose nuclei radiate atomic particles and energy

Recrystallization. Alteration of rocks whereby preexisting mineral grains are destroyed and new ones are formed, generally by increased heat and pressure; one of the metamorphic processes

Residuum. Soil formed in place by the disintegration and decomposition of rocks and the consequent weathering of the mineral materials

Sandstone. Sedimentary rock composed of sand grains

Saprolite. A soft, earthy, clay-rich, thoroughly decomposed rock formed in place by chemical weathering of igneous or metamorphic rocks

Scarp. Cliff or steep break in a slope

Schist. A visibly crystalline metamorphic rock containing abundant mica or other cleavable minerals so aligned that the rock breaks regularly along the mineral grains

Sedimentary. A term applied mainly to rocks formed of fragments of other rocks transported from their source and deposited in water; applies also to material transported in solution and deposited by chemical or organic agents

Sequence. A succession of stratified rocks

Shale. Platy sedimentary rock formed from mud or clay, breaking easily parallel to the bedding

Silica. Silicon dioxide (SiO_2), occurring as quartz and as a major part of many other minerals

Sill. A tabular body of igneous rock intruded along the bedding surfaces of stratified rocks

Sinkhole. A large depression caused by collapse of the ground into an underlying limestone cavern

Slickensides. Grooves or scratches in rocks made by movement along a fault surface

Slide scar. A term used to describe the resulting surface following a landslide

Slip. The amount of movement on a fault measured on the fault surface: strike slip is the component of slip measured along the strike of the fault: dip slip is the component measured in the direction of the dip of the fault

Solifluction. The slow flowage from higher to lower ground of water-saturated masses of waste

Sorted patterned ground. Polygons, stripes, and other patterns of broken rock material rearranged by repeated frost action

Strata. Beds or layers of sedimentary rock

Stratified. A term applied to rocks deposited in nearly horizontal layers of strata on the earth's surface

Strike. The direction or bearing of a horizontal line on a sloping bed, fault, or other rock surface

Syncline. A fold in stratified rocks in which the strata on opposite sides usually dip inward toward each other

Synclinorium. A large syncline composed of smaller synclines and anticlines

Talus. Blocks of rock pried loose by frost wedging; usually at the base of a cliff or other steep slope

Talus slope. The landform underlain by the talus

Tectonic. Pertaining to the larger structural features of the earth's crust and the forces that have produced them

Terraced deposit. A deposit of alluvial origin, composed of pebbles and cobbles in a sand or silt matrix, which is found on a relatively flat, elongate surface along the side of a valley; the deposit is formerly the alluvial floor of the valley

Thrust fault. A fault, commonly of low dip, on which rocks have slid or have been pushed laterally over other rocks

Tundra. One of the level or undulating treeless plains characteristic of arctic regions, having a black muck soil and a permanently frozen subsoil

Turbidity current. A heavy mixture of sediment and water that flows along the sea bottom in response to gravity

Vein. A rock containing ore minerals; usually a tabular mass

Water table. The upper surface of groundwater, below which soil and rock are saturated

Window. A hole produced by erosion through a thrust fault, exposing the underlying rocks

References

Literature Cited

Clark, Michael G., Patrick T. Ryan, and Eric C. Drumm. 1987. Debris slides and debris flows on Anakeesta Ridge, Great Smoky Mountains National Park, Tennessee. In *Landslides of Eastern North America*. U.S. Geol. Survey Circ. 1008: 18–19.

Clark, Michael G., and Allen C. Torbett. 1987. Block fields, block slopes, and block streams in the Great Smoky Mountains National Park, North Carolina and Tennessee. In *Landslides of Eastern North America*. U.S. Geol. Survey Circ. 1008: 20–21.

Delcourt, P. A., and H. R. Delcourt. 1979. Late Pleistocene and Holocene distributional history of the deciduous forest in the southeastern United States. *Veröffentlichungen des Geobotanischen Institutes der ETH* 5: 79–107.

———. 1981. Vegetation maps for Eastern North America: 40,000 yr. BP to the present. In *Geobotany II*, ed. R. Romans, 123–166. New York: Plenum Press.

———. 1985. Dynamic Quaternary landscapes of East Tennessee: an integration of paleoecology, geomorphology, and archaeology, Field Trip 7. In *Field Trips in the Southern Appalachians*, SE–GSA 1985, Univ. of Tennessee, Dept. of Geol. Sci., Stud. in Geol. 9: 191–220.

King, P. B. 1964. *Geology of the Central Great Smoky Mountains, Tennessee*. U.S. Geol. Survey Prof. Paper 349–C. 148 pp.

King, P. B., R. B. Neuman, and J. B. Hadley. 1968. *Geology of the Great Smoky Mountains National Park, Tennessee and North Carolina*. U.S. Geol. Survey Prof. Paper 587. 23 pp.

King, P. B., and A. Stupka. 1950. The Great Smoky Mountains— their geology and natural history. *Sci. Monthly* 71: 31–43.

Moore, H. L. 1978. Geologic and physiographic block diagram of East Tennessee. Tenn. Dept. of Transportation Geotech. Sect. files, Knoxville, TN.

———. 1986. Wedge failures along Tennessee highways in the Appalachian region: their occurrence and correction. *Bull. Assoc. Engineering Geologists* 23 (4): 441–60.

National Park Service. 1979. *Draft Environmental Impact Statement for General Management Plan, Great Smoky Mountains National Park*. Washington, D.C.: U.S. Dept. of Interior. 270 pp.

Neuman, R. B., and W. H. Nelson. 1965. *Geology of the western part of the Great Smoky Mountains, Tennessee*. U.S. Geol. Survey Prof. Paper 349–D. 81 pp.

Shafer, D. S. 1984. Late-Quaternary paleoecologic, geomorphic, and paleoclimatic history of Flat Laurel Gap, Blue Ridge Mountains, North Carolina. M.S. thesis, Univ. of Tennessee, Knoxville. 148 pp.

U.S. Geological Survey. 1949. *Topographic map of the Great Smoky Mountains National Park and vicinity*.

Additional Reading

Nontechnical

Albright, John. 1974. *Historic overview, Great Smoky Mountains National Park*. Denver: U.S. Dept. of Interior, National Park Service.

Frome, Michael. 1980. *Strangers in high places: the story of the Great Smoky Mountains*. Knoxville: Univ. of Tennessee Press. 391 pp.

Murlless, Richard, and Constance Stallings. 1973. *Hikers guide to the Smokies*. San Francisco: Sierra Club. 375 pp.

Technical

Bogucki, D. J. 1970. Debris slides and related flood damage associated with the September 1, 1951, cloudburst in the Mt. Le Conte–Sugarland Mountain area, Great Smoky Mountains National Park. Ph.D. diss., Univ. of Tennessee, Knoxville. 165 pp.

Carroll, Dorothy, R. B. Neuman, and H. W. Jaffe. 1957. Heavy minerals in arenaceous beds in parts of the Ocoee Series, Great Smoky Mountains, Tennessee. *Am. J. Sci.* 255 (3): 176–193.

Espenshade, G. H. 1943. Geology of the Hazel Creek Copper Mine area, Swain County, North Carolina. U.S. Geol. Survey open file report.

Gibbons, J. H., and Daniel Hale. 1956. Down into Bull Cave. *Natl. Speleological Soc. News* 14: 62–63.

Goldsmith, Richard, and J. B. Hadley. 1955. Pre-Ocoee erosion surface in the Great Smoky Mountains, North Carolina (abstract). *Geol. Soc. America Bull.* 66: 1687–1688.

Hadley, J. B., P. B. King, R. B. Neuman, and Richard Goldsmith. 1955. Outline of the geology of the Great Smoky Mountains area, Tennessee and North Carolina. In *Guides to Southeastern geology,* ed. R. J. Russell, 390–427.

Harris, Ann G., and Esther Tuttle. 1975. *Geology of National Parks.* Dubuque, Iowa: Kendall/Hunt Pub. Co. 554 pp.

Keith, Arthur. 1895. *Description of the Knoxville Sheet (Tennessee-North Carolina).* U.S. Geol. Survey Geol. Atlas, Folio 16. 6 pp.

King, P. B., J. B. Hadley, R. B. Neuman, and W. B. Hamilton. 1958. Stratigraphy of the Ocoee Series, Great Smoky Mountains, Tennessee and North Carolina. *Geol. Soc. America Bull.* 69 (8): 947–966.

King, P. B., J. B. Hadley, and R. B. Neuman. 1952. Guidebook of Excursion in Great Smoky Mountains, November 1–2, 1952. *Carolina Geol. Soc. Guidebook.* 60 pp.

Laurence, R. A., and A. R. Palmer. 1963. *Age of the Murray Shale and Hesse Quartzite on Chilhowee Mountain, Blount County, Tennessee.* U.S. Geol. Survey Prof. Paper 475–C.

Mathews, Raymond C., et al. 1975. Impact of Anakeesta Formation leachate, mineralized components and pH on the Shovel-nosed Salamander (*Leurognathus Marmoratus* Moore) of the Great Smoky Mountains National Park. *Assoc. of Southern Biologists Bull.* 22 (2): 68.

McMaster, W. M., and E. F. Hubbard. 1970. *Water resources of the Great Smoky Mountains National Park, Tennessee and North Carolina.* U. S. Geol. Survey Hydrologic Atlas HA–420.

Michalek, D. D. 1968. Fanlike features and related periglacial phenomena of the southern Blue Ridge. Ph.D. diss., Univ. of North Carolina, Chapel Hill. 198 pp.

National Park Service. 1974. *Draft Environmental Impact Statement, Wilderness Recommendation, Great Smoky Mountains National Park.* Washington, D.C.: U.S. Dept. of Interior. 270 pp.

Neuman, R. B. 1951. The Great Smoky Fault. *Am. J. Sci.* 249 (10): 740–754.

Neuman, R. B. 1947. Notes on the geology of Cades Cove, Great Smoky Mountains National Park, Tennessee. *Tenn. Acad. Sci. J.* 22: 167–172.

Reeder, Howard F. 1973. Sediment resulting from construction of Interstate Highway I-40 in North Carolina. U.S. Geol. Survey, open file report, Raleigh, NC.

Reheis, M. J. 1972. Periglacial features in the Southern Appalachian Mountains. MS. thesis, Univ. of Georgia, Athens. 26 pp.

Richter, D. M. 1973. Periglacial features in the central Great Smoky Mountains. Ph.D. diss., Univ. of Georgia, Athens. 148 pp.

Rodgers, John. 1953. Geologic map of East Tennessee with explanatory text. *Tenn. Div. Geol. Bull.* 58, Pt. 2.

Soil Conservation Service. 1970. *General soils maps and interpretation for Graham, Cherokee, Jackson, Macon, Clay, Haywood, and Swain Counties. North Carolina.* Washington, D.C.: U.S. Dept. of Agriculture.

Univ. of Tennessee. 1973. *Field conference guide for an altitudinal transect of the Great Smoky Mountains National Park.* 1973 International Geobotany Conference, Knoxville, TN.

White, S. E. 1976. Rock glaciers and block fields, review and new data. *Quaternary Research* 6: 77–97.

Wilson, C. W., Jr. 1935. The Great Smoky Thrust Fault in the vicinity of Tuckaleechee, Wear, and Cades Cove, Blount and Sevier Counties, Tennessee. *Tenn. Acad. Sci. J.* 10: 57–63.

Index

References to illustrations are printed in boldface type.